GPRA Project

GPRA:REPMES

Fractal Scale Cosmology

Framework Summary

Robert L. DeMelo

Fractal Scale Cosmology Framework Summary
Version 1.0.0.7, GPRA:RR 2.5.0.0
GPRA Project Book Reference Code: GPRA:REPMES:FSC
ISBN: 978-0-9810242-5-7
Copyright © March 2013, Robert L. DeMelo

15 Sagres Crescent, Toronto, Ontario, M6N 5E4
phone: 416-459-1500
email: r.demelo@gigaframe.com, mainframeii@gmail.com
website: www.gpofr.com, www.gpraproject.com

International Standard Book Number: 978-0-9810242-5-7
First Printed: 05-03-2013
Published in Canada

Trademarks
All terms mentioned in this book that are known to be trademarks or service marks have been appropriately capitalized. The publisher cannot attest to the accuracy of this information. Use of a term in this book should not be regarded as affecting the validity of any trademark or service mark.

Warning and Disclaimer
Every effort has been made to make this book as complete and as accurate as possible, but no warranty or fitness is implied. The information provided is on an "as is" basis. The author and the publisher shall have neither liability nor responsibility to any person or entity with respect to any loss or damages arising from the information contained in this book.

No Unlawful or Prohibited Use
As a condition of your use of the GPRA:REPMES:FSC book, you warrant that you will not use the GPRA:REPMES:FSC and its content for any purpose that is unlawful or prohibited by these terms, conditions, and notices. You may not use any content, application, device or derivative described in the GPRA:REPMES:FSC in any manner which could damage, disable, overburden, impair, injure or cause death to anyone. You may not also interfere with any other party's use and enjoyment of the information found in the GPRA:REPMES:FSC unless their use is capable of causing physical harm, injury or death to any other party. You may not obtain or attempt to obtain any materials or information or attempt to construct through any legal or illegal means any device capable of causing physical harm, injury or death to any other party with the information provided in the GPRA:REPMES:FSC. The information provided in the GPRA:REPMES:FSC is for educational purposes only. If any of these terms are not strictly adhered to, offenders will be punished to the full extent of the law.

Table of Contents

1   Abstract

Here a concise guide to the current state of fractal scale cosmology is given. It describes a unique fractal and velocity dependent scaling framework on quantum and macroscopic cosmology detailing an intrinsic, self-similar correlation to the geometry and material structure fundamentally constituting those cosmologies and respective dynamics. This highly predictive, heuristically formulated framework derives a series of interesting and precise results linking cosmic macroscopic objects to quantum scaled counterparts, and vice-versa. These results include deriving the elementary charge from its macroscopic scale relative, scaling planetary masses to quantum particle charges, describing the explicit relationship between charge and inertial mass of quantum particles, and demonstrating links through scaling between well known physical properties and constants. It further describes a consistent deterministic and mathematical explanation for matter, inertial mass and energy through interaction with a permeating spatial quantum medium (QM) constituted of sub-quantum particles, and subsequently constituted of smaller discretely scaled self-similar mediums and particles by scale factors of S forming a collective quantum medium (CQM). It gives an explicit mechanistic and mathematical explanation for gravity and orbital dynamics, subsequently deriving 43.6 (+/-0.4147) arc-secs for the precession of Mercury through computer simulation using the unique transform equations of this framework. The framework described in this document is continually evolving in order to reconcile all disparaging differences between macroscopic and quantum cosmologies. This research is described in elaborate detail, including simulation source code, in the book *"GPRA: Relativity and the Electrodynamics on the Phenomena of Matter and Energy through Scale (GPRA:REPMES)"*[1].

2 Research

Current research is in fractal scaling, discrete, self-similar cosmology and dynamics using the explicit framework described in this paper, which has been slowly evolved based on a simply derived scale factor S between macroscopic and quantum systems (between star systems and atomic systems) and its subsequent transformation equations since 2007. The hypothesis is that macroscopic and quantum cosmological systems are self-similar by a discrete scale factor of S and a significant difference in the passage of time of factor T. These differences in scale and time, if indeed true, signify that the Universe (the space and time continuum) is intrinsically fractal, materialistically and dynamically, similar to the Koch Curve and Mandelbrot Set. In a more complex geometric pattern distribution of scaled forms of matter using multiple scaling factors, the Universe may be more precisely a multifractal system ranging from smaller scales to the self-similar scale of S between the macroscopic and quantum.

This research is not simply the exploration or falsification of this initial hypothesis, but is an attempt to resolve all disparaging differences between the two scales through a highly objective and critical scientific analysis, if possible, in order to develop a better framework for continual testing and evolution. The process of resolving remaining disparaging differences between the two scaled cosmologies is in itself a process in falsification, which ultimately will invalidate or validate the hypothesis through probability. Currently the framework has evolved into a fairly robust predictive platform, while never losing touch with its simple foundations and physical context, by deriving many undeniably fascinating results through mathematics and computer simulation. These results indicate that this framework is definitely on the right path, and a possibility in uniting relativistic astrophysics and quantum physics through a uniquely defined scale projection.

Ultimately, the fractal scale cosmology of this framework maybe proven invalid, but as a scientist the importance lies in exploring all possibilities in order to unravel the truth. If the truth is that the Universe and the space-time continuum are not fractal, then that is just as important as finding that it is. As of right now, the probability is that this framework is indeed partially correct when considering its results. And the research continues. The ultimate goal of this research is to unravel all *truly core, irreducible and fundamentally elementary invariant laws* of physical reality, whatever they ultimately maybe. These core invariant laws would be the irreducible and most elementary building blocks of all other physical construct, including the dynamics and structure, that can be applied to all spaces (space-time medium, different inertial frames, quantum reference frame) through any translation. From these core laws, in an iterative series sum, all other contemporary physical law can be reproduced to our "current" level of knowledge, but also continually summed to reveal and explain larger complex patterns in nature like life.

3 Hypothesis

Cosmological and quantum systems, star systems and atoms, are a fractal self-similar scale-invariant representation of each other through a scaling factor (S) between the two systems. Universal matter is infinitely self-similar by a specific scale factor interval (S).

4 Important Insight

- After Pluto's 2006 demotion, realization of 4 rock planets (inner system) and 4 gas-giants (outer system) in Solar System

- If 4 is the cosmological atomic number, than the matching atom is Beryllium (Be)

- 5 major inner system objects (including Earth's Moon) + 4 major outer system objects = atomic weight of 9 (stable Be system)

5 Values Derived

The values derived throughout this document are either very precise or very close to many well known physical values. The following provides a summary list of these values derived later in this document:

Value	Symbol	How
Electron charge	q_e	Mass transform from gas-planet masses
Electron inertial mass	m_e	Relation to cross-sectional area of electron (gas-giant)
Neutron charge	q_n	Mass transform from rock-planet masses
Proton charge	q_p	Mass transform from fractional star masses
Planck mass	M_P	Relation to quantum nucleus (inner star system)
Planck length	L_P	Relation to quantum nucleus (inner star system)
Planck constant	h	Relation to photon charge (asteroid(s) mass)
Avogadro number	N_A	Time and scale transforms applied on ratio of electric and gravity constants
Precession of Mercury	Φ_{Merc}	Velocity dependent mass transform

6 Self-Similar Specifications

Sⁿ = fractal discrete scale factor

S^n = fractal discrete scale factor

Macroscopic Cosmology (S^0)		Type	Quantum Cosmology (S^{-1})	
	Star systems Inner system Outer system	System F-D Stats v < c	Atoms Nucleus Electron energy bands	
Inner System = Nucleus **From Star to Asteroid (inner) Belt**				
	Small stars, 1/4 our Sun	Baryon Fermion (large) v < c	Protons Form majority of nucleus	
	Rock planets (inner system) – inner system objects	Baryon Fermion (small) v > c	Neutrons (inner system) – no measurable charge dictates small macroscopic object relative; cross-sectional area of atomic nucleus includes orbiting neutrons adding to nucleus inertial mass	
Outer System = Energy Bands **Between Asteroid (inner) and Kepler (outer) Belts + Scattered Disk**				
	Asteroid Belt	Inner System Boundary v = c	Reservoir of Photons/Energy Lowest Energy Level	
	Gas-giant planets – large planets between Asteroid and Kuiper belts; valence planets are Uranus and Neptune	**Lepton Fermion (large)** v < c	Electrons – exist in atomic energy bands including valence electrons	
	Gas-giant moon	Lepton Fermion (small) v < c	Electron Neutrino	
Exo-System Objects				
	Asteroids, loosely bound space matter and dust (low density) - travelling in large wave like pattern balancing of gravity and	Gauge Boson (small) v = c	Photons, small gauge bosons – loose material object due to wave/pulse energy in quantum medium with ability to split/disperse itself; increased passage of	

	centrifugal forces, with decay and reconstitution over time along travel path; expelled and absorbed by asteroid belts in star systems		time aids particle-wave duality as all dynamical events are concentrated
	Nebulae	Gauge Boson (large) v < c	Large gauge bosons (W, Z, Higgs, and other unknowns)	
	Interstellar medium (ISM) + intergalactic medium (IGM) = cosmic medium (CM)	Gauge Boson (everywhere) wave v = c	Quantum medium (QM) - permeates known Universe; constituted of sub-quantum particles (S^{-2}); material scalar field; similarity to Higgs field but is a fractal constructive sum of self-similar, discretely scaled representations of itself, the collective quantum medium (CQM), and is materially deterministic; propagator of energy	

The following table summarizes the *four fundamental forces* which will be described in more detail later on throughout this book:

Force	Strength	Description	Macroscopic
Strong	1	Holds a nucleus together against the repulsion of same charge type protons. Exists at very short range 10^{-15} m (size of nucleus). Carrier particle is the gluon	**Cosmological scale relative to strong force is the star's own matter (material glue - gluons) binding cosmological protons together forming a visibly singular object** (ex. Sun with 4 cosmological protons). Space matter/dust obstructs and dampens the topological repulsion force allowing cosmological protons to attract and bind more readily. $S \times 10^{-15}$ m = roughly size of Sun.
Weak	10^{-6}	Responsible for quantum particle decay (ex. beta decay). Exists at a range of around 10^{-18} m Carrier particles are W and Z bosons with spin of 1	Deceleration, of planet (rock planet = cosmological neutron) by some obstruction, or change in surrounding quantum medium density, causes the planet to increase in size (absorbing quantum medium), destabilizing it and eventually leading to its decay by ejecting matter forming smaller constituents. Example of cosmological beta decay has a rock planet (neutron) decelerate into something larger than a cosmological proton (small nebula) that decays into a cosmological proton, gas-giant and gas-giant moon (or similar object). Acceleration is also included in this, **therefore cosmological weak force scale relative is a change in kinetic velocity, due to the**

Force	Strength	Description	Macroscopic
			quantum medium density change, that destabilizes the cosmological body resulting in it fragmenting. This deceleration of a star trans-mutates it into a type nebula from which the nebula coagulates back into other particles.
Electromagnetic	1/137	Holds atoms and molecules together. Range is infinite. Carrier particle is the photon with 0 *apparent* inertial mass and spin of 1. No known carrier particle for electric/electrostatic force.	Is a scale relative to the cosmological gravity force. The two forces derive from the same mechanism and dynamics, where electromagnetic force derives from sub-quantum medium (SQM) and gravity derives from quantum medium (QM). They are fractal scale relatives to each other. Force derives from a density-pressure differential in the respective mediums. The mediums are different because their constituents particles are different by a scale factor of S. QM does not penetrate fermion quantum particles, thus causes pressure on on quantum particles, and SQM does not penetrate fermion sub-quantum particles. This scale value, thus the mediums themselves, are dependent on their kinetic velocity in order to exist at different scales levels otherwise they would be the same medium. The constituents of the quantum medium exists at the kinetic velocity of $(17314.52)^4$ [m/s]. The constituents of the sub-quantum medium exist at the kinetic velocity of $(17314.52)^8$ [m/s].
Gravity	6x10^{-39}	Hold star systems, star system clusters and galaxies together. Range is infinite. No known identified carrier particle.	Same explanation for Electromagnetic force. Gravity equals electromagnetism (G = EM)[12] in mechanism and dynamic through velocity dependent discrete fractal scale interval of S. Gravity is a density-pressure differential imposed by the quantum medium (QM).

6.1 Relations Summary

- star systems (S^0) → atoms (S^{-1})

- gas-giant planets (S^0) → electrons (S^{-1})

- rock planets (S^0) → neutrons (S^{-1})

- stars (S^0) → one or multiple protons (S^{-1})

- inner star system (S^0) → atomic nucleus (S^{-1})

- outer star system (S^0) → electron energy levels of atom (S^{-1})

- # of rock planets + # of gas-giants → atomic mass (isotope #)

- # of gas-giants → atomic number

- groups of asteroids (one or many) + space dust (S^0) → photon (boson) (S^1)

- ISM/IGM (or cosmic medium) asteroids matter-wave velocity is c_s = 17314.5158 [m/s] (square root of) → c = 299792458 [m/s] for photons

- Asteroid Belt average velocity is 17314.5158 [m/s] → lowest atomic energy band at 299792458 [m/s]

- black holes, massive dense stars → large atomic nuclei like Uranium

- nebulae → large bosons (W, Z, Higgs, larger unknown bosons) → QM is itself an extremely large boson of very low charge density

- galaxies → macroscopic matter clump, large group of atoms (different types), or large molecule in empty space

- **cosmic medium (CM) (S^0)** = interstellar + intergalactic mediums (S^0) → **quantum medium (QM) (S^1)**

- ≈25% of CM particles are neutral, penetrate star system's heliosphere → ≈25% of QM penetrate atoms

- QM is comprised of sub-quantum particles (S^2), scale relative to CM matter (ISM/IGM plasma and particles) (S^0)

- **sub-quantum medium (SQM) (S^2)** is comprised of sub-sub-quantum particles (S^3), scale relative to CM

- **sub-sub-quantum medium (SSQM) (S^3)** is comprised of sub-sub-sub-quantum particles (S^4)

- sum of all fractal self-similar quantum medium scale relatives = **collective quantum medium (CQM)**

- there is a symmetry between natural object/particle types with pairs of large and small

7 Dynamics Specifications Summary

- QM highly penetrates macroscopic matter objects

- QM does not penetrate fermion quantum particles (Pauli Exclusion principle for QM)

- fermion quantum particles distort the QM when in motion → produce **magnetic field (or distortion field)**

- QM *almost* does not penetrate atomic nuclei (inner atomic system)

- QM does partially penetrate the outer atomic system (electron energy levels) by roughly 25%

- QM penetration into the atomic system is inversely proportional to velocity of atom

- QM density changes affect macro matter objects asymmetrically (matter density, size, scale)

- SQM density changes affect macro matter objects less asymmetrically (matter density, size, scale)

- CQM density changes can affect macro matter objects symmetrically (matter density, size, scale); *truly scale-invariant*

- QM, SQM, CQM experience dragging when fermion matter pushes against it → **frame dragging**

- moving matter object distorts QM forming a distortion field and *localized* QM dragging

- localized QM dragging was not considered by Michelson-Morley differing QM from classical ether

- SQM drags more slowly (is more stationary) than QM, and SSQM is even more stationary; solar wind plasma moves faster than QM

- QM, SQM, CQM densities and pressure are proportional to object's velocity from object's perspective

- low matter density QM achieved by scattering (high kinetic energy) using quantum particles = **hot** space

- high matter density QM (less scattered, high potential energy, deep space) = **cold** space

- varying QM density refracts photon path (curves it)

- low density QM **expands** photon boson wave size **slowing** it but **permits** faster fermion particle movement

- QM imposes **drag** on fermion matter proportional to QM density, which equates to inertial mass (in acceleration or through pressure)

- QM drag coefficient **Cd** on fermion matter approaching c increases matching closely Lorentz factor

- QM density changes affect relative experience of distance and passage of time → QM is classical space-time medium (ζ)

- Higgs field is similar to the QM and CQM fractal construct, but is **limited/void** of fractal topology, and derivative relations

- Dark Matter is a derivative of cool denser QM; Dark Energy is a derivative of hotter less dense QM (hotter than Dark Matter)

- hot space **permits** faster fermion/macro matter movement (refraction, time dilation)

- cold space **restricts** fermion/macro matter movement (velocity time dilation)

- hot space **melts (phase transition)** slower moving objects (phase transition, time dilation)

- hot space **phase transition effect** is counterbalanced by increasing object velocity

- cold space **expands/grows** *slow* moving macro-matter objects (decreasing velocity); QM absorbed

- all matter objects are continuously expelling QM/CQM matter back into the QM/CQM environment

- slow moving objects produce less magnetic/distortion shielding allowing more QM/CQM to bind to object (aiding size and mass increase)

- fast moving objects produce more magnetic shielding prevent QM/CQM binding to object

- velocity changes by QM drag can change planets types into other planet types (link to *Expanding Earth Theory*)

- QM matter binds to slow moving fermion/macro objects due to reduced magnetic field

- cold space **expansion/growth** of macro-matter objects is counterbalanced by increasing velocity

- macro-matter objects **contracts/shrinks** by *increasing in velocity* (relative QM density increases - cools)

- higher velocity fermion/macro objects **release** more sub-quantum matter back into QM than absorb

- gravitational potential energy inversely proportional to QM density

- electric potential energy inversely proportional to SQM density (*scale relative to gravitation potential*)

7.1 Cosmological Scale Level vs. Velocity Map

medium (acronym)	CM	QM	SQM	SSQM	SSSQM	...	nSQM
level (n)	0	1	2	3	4	...	n
scale (factors of S)	S^0	S^{-1}	S^{-2}	S^{-3}	S^{-4}	...	S^{-n}
wave velocity [m/s]	$c_o^{0.5}$	c_o^1	c_o^2	c_o^4	c_o^8	...	$c_o^{((2^n)/2)}$
constituent kinetic velocity [m/s]	c_o^1	c_o^2	c_o^4	c_o^8	c_o^{16}	...	$c_o^{(2^n)}$

Table 1: Cosmological scale level vs. velocity map. Here c_o=299792458 and its square root is c_s=17314.5158 are numeric values only, where the resulting product is a velocity [m/s] in the map. The factors of scale (S) is a ratioed velocity function $s=(v'/v_o)^e$ explained later, where at $v'= c_o$ [m/s] and $v^o = 1$ [m/s] then s=S. Sub-sub quantum medium (SQM) and sub-sub-sub-quantum medium (SSQM) wave velocities are ratioed with their initial gravitational terminal/wave velocity to produce their respective scale factor. Cosmic medium (CM)'s interstellar velocity can be between 1 [m/s] to c_s [m/s] changing $c_s^{0.5}$ to c_s^o. CM constituent kinetic velocity is the speed of light c_o [m/s] which is directly related to $E=mc^2$. Sub-quantum medium (SQM) wave velocity is 8.9875x10^16 [m/s], which means a SQM wave can travel to Alpha Centauri, which is 4.1343x10^16 [m] (4.37 [light years]) away, in 0.46 [seconds] which may explain quantum entanglement.

8 Mathematical Specifications

The mathematical specifications of this framework were developed over many years. The core framework is detailed in this section.

8.1 Discrete Scale Factor

The scale difference between the Solar System (Sol) and the Beryllium (Be) atom is the average distance between the Kuiper belt and scattered disc (77.5 [AU]) in ratio to the empirically measured radius of the Beryllium atom (105 [pm]), or the average radius between covalent (96±3 [pm]) and calculated radii (112 [pm]):

$$S=\frac{r_{sol}}{r_{Be}}=\frac{77.5[AU]}{105[pm]}=1.1025 \times 10^{23} \tag{1}$$

It was subsequently found that S related to numerical value of speed of light c and mathematical constant e *(the **fractal dimension** between macro and quantum)*:

$$S=(299792458)^e=1.1025 \times 10^{23} \tag{2}$$

8.2 Relative Scale Function

Factor S can be formulated as a relative velocity ratio between c and 1 [m/s]:

$$S=\left(\frac{c}{1[m/s]}\right)^e=\left(\frac{299792458[m/s]}{1[m/s]}\right)^{2.71828\ldots}=1.1025 \times 10^{23} \tag{3}$$

Using velocity v in place of c in equation (3), a ratio between velocities states in ratio to 1 [m/s] is:

$$y=\frac{\left(\frac{v}{1[m/s]}\right)}{\left(\frac{v_o}{1[m/s]}\right)}=\left(\frac{v}{v_o}\right)\approx\sqrt{\frac{K}{K_o}}=\sqrt{\frac{\frac{1}{2}mv^2}{\frac{1}{2}mv_o^2}} \tag{4}$$

$$\frac{v}{v_o}=\frac{\Delta v+v_o}{v_o}=1+\frac{\Delta v}{v_o}=1+\frac{\Delta v}{v_o+v_r}$$

Where v_r is the rotational velocity at the planet's surface, which is more effectively dragging the quantum medium than at distances further away forming *localized* frame dragging, v_o is the initial velocity state, v is the resulting velocity state. Therefore, the S scale factor is seemingly a function of velocity. When $v = c$ the function $s(v)$ results in S:

$$s(v) = \left(\frac{v}{v_o}\right)^e = (\gamma)^e$$

(5)

Resulting in a simple length transform equation of:

$$l = \frac{l_o}{S} = \frac{l_o}{s(v)} = \frac{l_o}{\left(\dfrac{v}{v_o}\right)^e}$$

(6)

Extending equation (4) to include an initial velocity of an orbiting planet at the macroscopic scale so that of the velocity of the quantum scale relative in an atom gives the following:

$$\gamma = \left(\frac{v}{v_o}\right)^\alpha = \left(\frac{v}{v_o}\right)^{\left(2 - \log_{\left(\frac{c}{c}\right)}\left(\frac{v}{c}\right)\right)}$$

(7)

8.3 Quantum Medium Dependence

The velocities used are naturally set by the physical dynamics of the Universe through gravitation, therefore these velocities, v and v_o, are dependent on the quantum medium (QM) density ρ_{qm}, and are essentially natural terminal velocities v_t for matter objects traversing through the quantum medium (QM), which is dependent on the mass distribution of macroscopic matter. The terminal velocity due to quantum medium is inversely proportional the matter density of the quantum medium ρ_{qm} at point d_o, Where d_o and d_i are Cartesian points.

$$v_t = \sum_{i=0}^{n} \sqrt{\frac{Gm_i}{|d_o - d_i|}} + v_\wedge \propto \frac{1}{\rho_{qm}(d_o)}$$

(8)

Where v_A is a reasoned constant, underlining terminal velocity of space void of mass objects (where all $m_i = 0$). Velocities that deviate from the terminal velocity (v_t) in a particular region of space experience stress forces. A net velocity can be derived from total kinetic energy using the terminal orbital and rotational velocities of a planet (at 90°):

$$v_{net} = \sqrt{v_{orb}^2 + v_{rot}^2}$$

(9)

The force experienced is gravitational force from the potential energy from quantum medium density gradient:

$$F_g = -\nabla U$$

(10)

For a planet orbiting a star, the orbital velocity is a product of gravity, but is also a terminal velocity through the quantum medium as the star's rotating interplanetary magnetic field (IMF) produces an irrotational vortex of solar wind (heliospheric magnetic current sheet) pushing the planet in its orbital direction and in balance with centripetal force due to gravity; the quantum medium density-pressure differential pushing inward and quantum medium's drag terminates the orbital velocity of the planet. The system of forces is in balance.

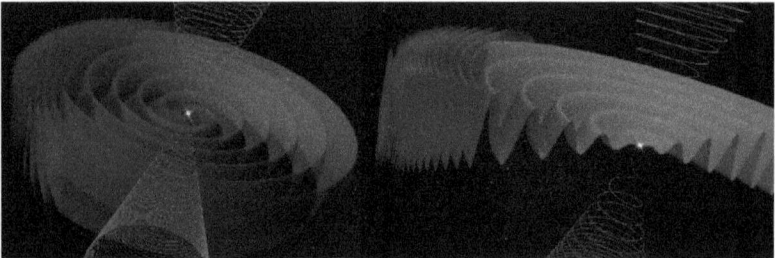

Figure 1: Interplanetary magnetic field (IMF) solar wind, represented here by the rotating heliosphere magnet current sheet, push the planets along their orbits. Image from NASA Goddard Space Flight Center Heliosphysics.

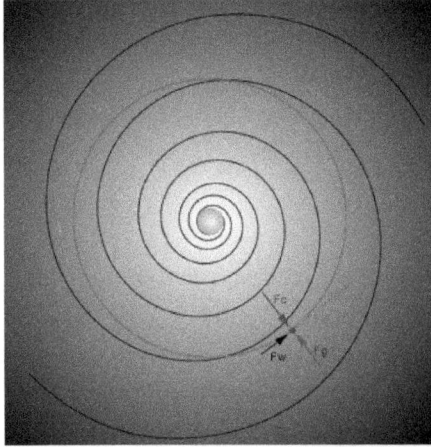

Figure 2: solar wind force F_w pushes planet causing linear motion and centrifugal force F_c. Quantum medium (QM) density-pressure gradient, where white region is less dense QM than dark region, pushes planet inward F_g counter-balancing centripetal force F_c. Quantum medium drag force F_d counter-balances the solar wind force F_w resulting in terminal orbital velocity v_t.

Where

$$F_{solar\,wind} = F_w = F_d = F_{QM\,drag}$$
$$F_{centripetal} = F_c = F_g = F_{gravitation} = F_{QM\,pressure} \tag{11}$$

Therefore velocity from gravitation, terminal velocity from the quantum medium, and centripetal velocity are in balance:

$$F_c = F_g = m\frac{v_c^2}{d} = \frac{GMm}{d^2} \quad \rightarrow \quad v_c = \sqrt{\frac{GM}{d}} = v_t \tag{12}$$

The terminal velocity (v_t) is a product of drag force (F_d) through the quantum medium density:

$$F_d = \frac{1}{2} C_d \rho_{qm} v_t^2 A \quad \rightarrow \quad v_t = \sqrt{\frac{2F_d}{C_d \rho_{qm} A}} \tag{13}$$

Where A is the summed cross-sectional area of the quantum constituents of macroscopic matter object.

Equation (13) is best defined for large objects traversing a gaseous medium, where Stokes law expresses terminal velocity for relatively slow moving small objects of density ρ_o and radius r_o traversing a gaseous medium, where μ is the dynamic viscosity [N s/ m²], consisting of constituents closer in size to the traversing objects inducing viscous drag:

$$v_t = \frac{2}{9\mu}(\rho_o - \rho_{qm})a r_o^{2}$$

(14)

Where a is the acceleration of the force applied on the planet by the solar wind's irrotational vortex. Both equations (13) and (14) are derivative formulations from the *Navier–Stokes equations*. For simplicity, drag force of equation (13) will be referenced primarily due to the relation with aerodynamics of larger objects. Both equations are related to the object's radius r_o squared.

The mass density of the quantum medium is different than its matter density but are proportional (to a point). Macroscopic matter objects scatter and dilute the quantum medium lowering its matter density near the surface radius of the macroscopic object, and less so the further away from the surface of the object. The scattering and dilution of the quantum medium is produced by the expelled energy quanta in the form of quantum particles, because the quantum medium does not penetrate fermion quantum particles. This can easily be seen by the energy released by the Sun.

The quantum medium density is inversely proportional the gravitational potential energy scalar field:

$$\rho_{qm} \propto \frac{1}{|U_g|} = \frac{d^2}{GM}$$

(15)

The quantum medium density ρ_{qm} is also inversely proportional to the luminosity flux density (Ψ_L) of the released energy quanta of the macroscopic matter object at distance d:

$$\rho_{qm} \propto \frac{1}{\Psi_L} = \frac{4\pi d^2}{L} = \frac{4\pi d^2}{\sigma A_o T^4} = \frac{4\pi d^2}{\sigma (4\pi r_o^2) T^4} = \frac{1}{\sigma T^4}\left(\frac{d^2}{r_o^2}\right)$$

(16)

Where L is luminosity, T here is temperature, d is distance from object, r is object radius, and σ is Stefan-Boltzman constant (5.670373×10⁻⁸ [W m⁻² K⁻⁴]). Luminosity as an estimated formula of in reference to the Sun:

$$\frac{L}{L_S} \approx \left(\frac{M}{M_S}\right)^{3.9} = \left(\frac{R}{R_o}\right)^2 \left(\frac{T}{T_o}\right)^4 \tag{17}$$

Where L_S and M_S are the luminosity and mass of Sun respectively.

The temperature T equation (16) can further be reduced to its relation with gaseous kinetic energy (of the expelled energy):

$$E_{K_i} = \frac{1}{2} k_i T = \frac{1}{2}\left(\frac{k_g}{N_A}\right) T \quad \rightarrow \quad T = \frac{2 E_{K_i} N_A}{k_g} \tag{18}$$

Where k_g is the ideal gas constant (8.3144621 [J K^{-1} mol^{-1}]) and N_A is Avogadro's number (6.0221415x10^{23} mol^{-1}). Combining equations (16) and (18) the following is achieved:

$$\rho_{qm} \propto \frac{1}{\psi_L} = \frac{k_g}{\sigma\left(2 E_{K_i} N_A\right)^4}\left(\frac{d^2}{r_o^2}\right) \tag{19}$$

Changes to the quantum medium density affect the perception of distance and passage of time by an observer in, and ouside, the changing density. This intricately binds the quantum medium to the space-time continuum where thermodynamics plays a significant role.

Combining the equations (15) and (19):

$$\frac{d^2}{GM} = k_L\left(\frac{k_g}{\sigma\left(2 E_{K_i} N_A\right)^4}\left(\frac{d^2}{r_o^2}\right)\right) \tag{20}$$

Where k_L is an unknown equating compensating constant. Using equation (20) we solve for terminal velocity giving a more fulsome representation of the dynamics affecting the terminal velocity of matter object through the quantum medium, and binding classical gravity formula to this more detailed underlining material dynamic of space:

$$v_t = \sqrt{\frac{GM}{d}} = \sqrt{\left(\frac{\sigma\left(2\,E_{K_i} N_A\right)^4}{k_L k_g}\left(\frac{r_o^2}{d}\right)\right)} = \sqrt{\frac{2\,F_d}{C_d \rho_{qm} A}} \tag{21}$$

Macroscopic rest mass of object causing gravitational potential in this new dynamic is simply derived from (21):

$$M = \frac{\sigma\left(2\,E_{K_i} N_A\right)^4 r_o^2}{G k_L k_g} = \left(\frac{16\,\sigma\,N_A^4}{G k_L k_g}\right) E_{K_i}^4 r_o^2 = (k_5)\,E_{K_i}^4 r_o^2 \approx \frac{E}{c^2} \tag{22}$$

Where k_5 is the product of five constants.

8.4   Length

Using equation (5) the following length transform equation is derived:

$$l = \frac{l_o}{s(v)} \quad \rightarrow \quad l = \frac{l_o}{S} \tag{23}$$

Where the radius of a cosmological proton using the proton charge radius (R_p = 0.8775(51)x10^{-15} [m]) is in between to the radius of the Jupiter (7.1492x10^7 [m]) and the Sun (6.9634x10^8 [m]):

$$R_o = R_p\,S = 9.6744 \times 10^7 \, [m] \tag{24}$$

8.5   Rest Mass

Combining equation (22) and (23) results in the following mass rest equation with scaling factored in:

$$M = \left(\frac{\sigma\left(2\,E_{K_i} N_A\right)^4}{G k_L k_g}\right)\left(\frac{r_o}{s(v)}\right)^2 = \frac{M_o}{s(v)^2} = \frac{M_o}{S^2} \tag{25}$$

8.6   Planck Relation

Applying this equation (23) to the radius of our Solar System and scaling it to the sub-quantum scale (S^{-2}) the following is achieved in comparison to Planck Length

(L_P) indicating Planck Length is a sub-quantum length of sub-quantum ion closely matching the scaled version of a cosmological Beryllium atom:

$$r = \frac{r_{sol}}{S^2} = \frac{77.5[AU]}{S^2} = 9.56 \times 10^{-34}[m]$$

$$\approx L_P = \sqrt{\frac{\hbar G}{c^3}} = 1.6162 \times 10^{-35}[m]$$

(26)

And if using the radius of the inner Solar System (system nuclei) up to Mars, the following is achieved:

$$r = \frac{r_{(inner\ sol)}}{S^2} = \frac{1.5237[AU]}{S^2} = 1.8803 \times 10^{-35}[m]$$

$$\approx L_P = 1.6162 \times 10^{-35}[m]$$

(27)

Where the macroscopic length of Planck Length (L_P) would be at two S scale intervals larger (S^2):

$$r = L_P S^2 = 1.6162 \times 10^{-35}[m] S^2 = 1.9645 \times 10^{11}[m] = 1.31[AU]$$

(28)

This radius may represent the average radii of a quantum atom or nuclei, comprised of sub-quantum particles.

The value in equation (28) is almost an average distance between Earth and Mars; an average boundary range of the inner system's furthest orbiting planets. Consider the following:

$$m \frac{v^2}{d} = \frac{GMm}{d^2} \rightarrow mv^2 = \frac{GMm}{d} \rightarrow M = \frac{d\, v^2}{G}$$

(29)

When velocity v in equation (29) is set to the speed of light c and d is the radius of the inner solar system (macroscopic Planck Length), Planck Mass (M_P) relates through mass transform scale factor S as so:

$$M = \frac{d\, c^2}{G} = \frac{(1.9645 \times 10^{11}[m]) c^2}{G} = 2.6471 \times 10^{38}[kg] \approx M_P S^2$$

(30)

Where Planck Mass (M_P) scaled using equation (25) is:

$$M_P = \sqrt{\frac{\hbar c}{G}} = 2.1765 \times 10^{-8}\,[kg] \quad \rightarrow \quad M_P S^2 = 2.6456 \times 10^{38}\,[kg] \tag{31}$$

This is an interesting coincidence, one of many using this framework, and also strengthens the scale factor S. It appears to relate once again to the nuclei of a system, or the Beryllium ion in our case. Some reasoning would state, if these values are boundary conditions, that a very small unknown particle (tiny black hole) of Planck Length can have the mass of Planck Mass. Similar reasoning through relative scale can state that an unknown macroscopic matter object may exist having the radius of the inner Solar System with the mass given in equation (30). Could these extremes be boundary limits in minimum particle size to maximum macroscopic mass? Why would the radius of cosmological Beryllium ion be common to both scenarios giving exact values? This may indicate that the physics experienced in our Solar System may be relative to our star system type, scale and relative velocity through the cosmos.

8.7 Inertial Mass

Inertial mass is the quantitative measure of a matter object's resistance to acceleration. This inherently describes that mass doesn't exist, or is unmeasurable, if no acceleration is applied. Understanding that the quantum medium (QM) does not penetrate fermion quantum particles or atomic nuclei (more so binds to them), we considering a small linearly moving macroscopic matter object of radius r with n atoms. Applying equation (23) to radius of the macroscopic object results in the premature density equation:

$$\rho = \frac{m}{\frac{4}{3}\pi r^3} = \frac{m}{\frac{4}{3}\pi \left(\frac{r_o}{s(v)}\right)^3} \tag{32}$$

Since the object is moving through the quantum medium (QM), the cross sectional area of its fermion quantum particles and atomic nuclei are proportional to the drag (F_d) imposed through the medium:

$$F_d = \frac{1}{2} C_d \rho_{qm} v^2 A = m a_d \tag{33}$$

24

Where a_d is the drag acceleration. Equation (33) gives a formulation for inertial mass which is proportional to cross-sectional area A and the mass density of the quantum medium ρ_{qm}. Applying the Equivalence Principle to the quantum particle being pressured into motion by a pushing quantum medium, the following is achieved:

$$m = \frac{C_d \rho_{qm} v^2 A}{2a} = \frac{1}{2} C_d \rho_{qm}(at^2) A \tag{34}$$

Where a is acceleration and t is the time under acceleration. When applied acceleration is null, inertial mass of an object cannot be measured though it maybe moving at a constant velocity. Macroscopic accelerations are relatively small compared to electric potential accelerations on a quantum particle. This means a macroscopic object maybe under applied acceleration and experience little change in inertial mass. Also quantum particles will achieve terminal velocity resulting in zero acceleration.

Applying equation (23) to the radius of the cross-sectional area results in the following:

$$A = \sum_{i=1}^{n} \pi r_i^2 = \sum_{i=1}^{n} \pi \left(\frac{r_{o_i}}{s(v)} \right)^2 = \frac{A_o}{s(v)^2} \tag{35}$$

Combining equations (34) and (35):

$$m = \frac{\left(\dfrac{C_d \rho_{qm} v^2 A_o}{2a_d} \right)}{s(v)^2} = \frac{m_o}{s(v)^2} \quad \rightarrow \quad m = \frac{m_o}{S^2} \tag{36}$$

The macroscopic mass of the object is the sum of its constituents (m_{oi}) which the quantum medium does not penetrate:

$$m_{macro} = \frac{\displaystyle\sum_{i=1}^{n} m_{o_i}}{s(v)^2} = \frac{m_{o_{macro}}}{s(v)^2} \quad \rightarrow \quad m_{macro} = \frac{m_{o_{macro}}}{S^2} \tag{37}$$

Equation (37) applied to Jupiter's mass transforms into the numerical value of an elementary electron charge by 97.45% similarity:

$$m=\frac{m_{Jupiter}}{S^2}=\frac{1.898\text{x}10^{27}[kg]}{S^2}=1.5614\text{x}10^{-19}[kg]$$

(38)

Subsequently using equation (5) by taking into accounting total kinetic velocity in the form of linear and angular velocities along with the mass and velocity values of Jupiter's moons, the value matched $1.6022\text{x}10^{-19}$. This supports the hypothesis that gas-giants are cosmological electrons. The significance of this can be summed up in a quote by Paul Dirac recounted by Neil Toruk in his latest book[2] in which Dirac states in similar context, *"physics would never make any progress until we understood how to predict the exact value of the electric charge carried by an electron."*

Subsequently all cosmological bodies have been mapped to their quantum counterparts where mass becomes charge value, as charge is derived from drag on quantum particles through the sub-quantum medium (*SQM*)[1]. This is the result that has been the prime motivator to continue this research.

Cosmological objects mapped to quantum counterparts using this framework:

obj	v net:c (ratio) m/s	v net:c c	vnet c qmass C (or kg)	qmass avg. C (or kg)	expected q C (or kg)	actual v:c m/s	actual v c
Sun	8.30E+08	2.77	6.47E-19		*6.41E-19*	8.31E+08	2.77
Mercury	8.29E+08	2.76	1.08E-25		*8.73E-24*	3.69E+08	1.23
Venus	6.06E+08	2.02	8.70E-24		*8.73E-24*	6.06E+08	2.02
Earth	5.16E+08	1.72	2.57E-23		*8.73E-24*	6.29E+08	2.1
Moon	5.16E+08	1.72	3.17E-25		*8.73E-24*	2.80E+08	0.93
Mars	4.17E+08	1.39	8.79E-24	8.73E-24	*8.73E-24*	4.17E+08	1.39
Jupiter	3.14E+08	1.05	1.21E-19		*1.60E-19*	2.98E+08	1
Saturn	2.39E+08	0.8	1.59E-19		*1.60E-19*	2.39E+08	0.8
Uranus	1.26E+08	0.42	7.90E-19		*1.60E-19*	1.69E+08	0.56
Neptune	1.05E+08	0.35	2.55E-18	9.04E-19	*1.60E-19*	1.74E+08	0.58

Table 2: Cosmological planets mapped to quantum particles at the respective ratioed velocity.

Supporting quantum transmutation, and Expanding Earth theory, as applied to all cosmological planets, the following is derived using the transforms equations of this framework:

Object	m (kg)	r (m)	v (m/s)	v' (m/s)	m' (kg)	r' (m)	Comp mass
Mars	6.400E+23	3.396E+06	24077	13070	1.773E+25	1.787E+07	Uranus
Earth	5.974E+24	6.378E+06	29780	13070	5.256E+26	5.982E+07	Saturn
Venus	4.868E+24	6.051E+06	35020	13070	1.034E+27	8.818E+07	Jupiter
Mercury	3.302E+23	2.439E+06	47870	13070	3.836E+26	8.313E+07	Saturn
Mars	6.400E+23	3.396E+06	24077	5430	2.102E+27	1.946E+08	Jupiter
Earth	5.974E+24	6.378E+06	29780	5430	6.231E+28	6.514E+08	x10 Jupiter
Venus	4.868E+24	6.051E+06	35020	5430	1.226E+29	9.601E+08	small star
Mercury	3.302E+23	2.439E+06	47870	5430	4.548E+28	9.051E+08	x10 Jupiter
Mars	6.400E+23	3.396E+06	24077	4666	4.793E+27	2.939E+08	Jupiter
Earth	5.974E+24	6.378E+06	29780	4666	1.421E+29	9.837E+08	small star
Venus	4.868E+24	6.051E+06	35020	4666	2.795E+29	1.450E+09	small star
Mercury	3.302E+23	2.439E+06	47870	4666	1.037E+29	1.367E+09	small star

Table 3: Rock planets transform to gas giants, stars, nebulae and decaying in density by changing their velocity to v'.

The densities are fairly low for some results in order to be truly considered a small star (indicated by mass), more justly it indicates the decay of the object into a nebulae at very low kinetic velocities (linear and rotational) as in the case of Mercury transitioning from velocity 47870 [m/s] to 4666 [m/s]. Again the velocity transition is based on a change in quantum medium density.

The assertion of this research, due to its findings, is that charge (Q) is fundamentally mass (M) derived by the same *invariant mechanism* (drag), though dependent on their respective spatial mediums; quantum medium (QM) for mass (M), sub-quantum medium (SQM) for charge (Q). This can be perplexing. To demonstrate dimensional similarity[12], a dimensional analysis of the gravitational constant and Coulomb's electric force constant is given:

$$D_{k_e} = N.m^2/C^2 = \frac{(kg)(m)(m)^2}{(s)^2(C)^2} \rightarrow \frac{(M)(L)^3}{(T)^2(Q)^2} \tag{39}$$

And

$$D_G = m^3/kg/s^2 = \frac{(m)^3}{(kg)(s)^2} \rightarrow \frac{(L)^3}{(M)(T)^2} \tag{40}$$

By converting charge (Q) to mass (M), the respective dimensions derive unity:

$$\frac{(M)(L)^3}{(T)^2(Q)^2} \rightarrow \frac{(L)^3}{(M)(T)^2}$$
$$\text{where } M=Q$$
$$D_{ke}=D_G$$
$$\frac{(L)^3}{(M)(T)^2}=\frac{(L)^3}{(M)(T)^2}$$

(41)

For convention on the transitioning of dimension, from mass or charge in this case, the following notation was devised:

$$[kg=C]$$
$$\text{where } m=157[kg]$$
$$\text{becomes } m=157[kg=C]$$

(42)

Where square brackets encase units representing dimension not variables. Also charge will be referred to as quantum mass, in contrast to its inertial mass. Inertial mass of quantum particles and the relation to its charge will be explicitly detailed later.

Therefore combining equations (37) to (32) the following density equation is produced:

$$\rho=\frac{m}{\frac{4}{3}\pi r^3}=\frac{\dfrac{m_o}{s(v)^2}}{\dfrac{4}{3}\pi\left(\dfrac{r_o}{s(v)}\right)^3}=s(v)\rho_o \quad \rightarrow \quad \rho=S\rho_o$$

(43)

The neutron's mass is that of a proton, yet its charge magnitude is essentially non-existent. This indicates a very small object. A quantum particle having the inertial mass near that of a proton, which has a positive elementary charge, while having no measurable charge in this framework is illogical if one omits how the cross-sectional area of the hypothesized quantum scale relative rock planet contributes to the inner system (nucleus). Rock planets contribute to the magnetic field and cross-sectional area of the inner system. This adds drag passing through the ISM/IGM (cosmic medium). In this framework, the fractal scale nucleus model has neutron's orbiting a proton-based quantum object (scale relative to a star). Their orbit contributes to the quantum medium distortion field around the nucleus adding drag and inertial mass as the collective system moves. The quantum medium still **partially** passes through the space between the proton-based core quantum object and its orbiting neutrons.

The following figure depicts how the cross-sectional area of the neutron is added to the nucleus cross-sectional area:

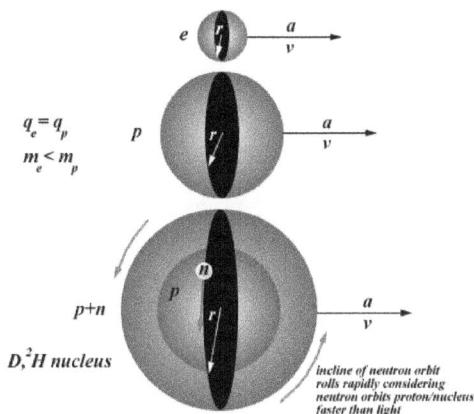

Figure 3: Inertial mass dependent on cross-sectional area

The orbital radius of a neutron (n) around a proton (p) in a Deuterium (D, ^2H) is :

$$\frac{m_{(p+n)}}{m_{(p)}} = \frac{k_f a_{A(p+n)} t^2 A_{(p+n)}}{k_f a_{A(p)} t^2 A_{(p)}}$$
$$= \frac{1.673 \times 10^{-27}[kg] + 1.675 \times 10^{-27}[kg]}{1.673 \times 10^{-27}[kg]} = 2.00119546 \tag{44}$$

If all parameters remain the same except the cross-sectional area the following ratio is true:

$$\frac{A_{(p+n)}}{A_{(p)}} = \frac{r_{p+n}^2}{r_p^2} = 2.00119546$$
$$r_{(p+n)} = (\sqrt{2.00119546})(r_p) = (1.4146)(0.8775 \times 10^{-15}[m]) \tag{45}$$
$$r_{(p+n)} = 1.7561 \times 10^{-15}[m]$$

At the macroscopic scale this radius would be:

$$r = S(r_{(p+n)}) = 1.9360 \times 10^{+8}[m] \tag{46}$$

29

The actual measured charge radius of Deuterium is 2.1402x10⁻¹⁵ [m] which is close to the calculated value of formulation (45). This gives a cosmological radius of 2.3596x10⁸ [m] which is comparable to the orbital radii of Jupiter's moons. It is also comparable to a small star radii or very large gas-giant. The measured charge radii of a proton is 0.8775x10⁻¹⁵ [m]. The cosmological radius of a proton is:

$$r = S(r_p) = S(0.8775 \times 10^{-15}[m]) = 9.6744 \times 10^7 [m]$$

(47)

This radii is comparable to the radius of Jupiter. Interestingly, Jupiter has 4 large moons comparable in size to Mercury, where 4 is the atomic number of the Beryllium system. This may indicate a hidden dynamical relation (or symmetry) where Jupiter, at the highest kinetic energy orbital for gas-giants, maybe a companion dwarf star (only at that orbital) common to most Beryllium system formations. If true, this relation would also exist at the atomic scale.

Considering that about 20%-25% of the ISM atoms are neutral, not ions, and are not deflected by the heliosphere and planetary magnetic fields (distortion fields). This same dynamic applies at the quantum scale with the atomic nuclei and quantum medium, which decreases the drag effect and subsequently the inertial mass. To compensate, the cross-sectional area of the orbiting neutron would have to increase by about ~5 times (100%/20%). Using the radius of the Sun, the orbit would reach 10⁹ [m] to 10¹⁰ [m] which is close to the orbital radius of Mercury.

Considering the planetary magnetospheres and their field strength, the magnetic field (a medium distortion field) of the rock planet (neutron) plays a significant role in shielding the rock planet and impeding the motion of interstellar medium particles passing through the system. Relatively through scale, the same dynamic occurs at the quantum scale. The scale relative to the magnetosphere of a rock planet is a sub-quantum medium distortion field produced by the motion and internal dynamics of the neutron. The Biot-Savart Law applies here where the field strength is proportional to the velocity of the object. Since rock planets travel at the system velocity *and* at their orbital velocities, their magnetic field strength is higher than a cosmological proton (star):

$$\vec{B} = \left(\frac{\mu_o}{4\pi}\right)\frac{q}{r^3}(\vec{v} \times \vec{r}) \quad \rightarrow \quad \vec{D} = \left(\frac{\mu_D}{4\pi}\right)\frac{M}{r^3}(\vec{v} \times \vec{r})$$

$$SQM \ derives \ B \quad \rightarrow \quad QM \ derives \ D$$

(48)

Where B is the magnetic field strength derived from the sub-quantum medium (SQM), μ_o is vacuum permeability equal to 4π×10−7 [V s/(A m)] ≈ 1.2566×10⁻⁶

[H·m^{-1}] or [N·A^{-2}] or [T·m/A], r is the distance the field measurement is taken away from the object, and q is the charge value of the object. Relatively through scale, D is the macroscopic equivalent to B called distortion field strength derived from the quantum medium (QM), and M is mass of the macroscopic object, and μ_D is an unknown constant similar, through scale, to μ_o the vacuum permeability. Another way to measure distortion strength is through *ratio* in the difference between time measurements in front and back of object in direction of motion (result is dimensionless). Distortion change the density of the quantum medium, and subsequently the density of collective quantum medium (CQM)

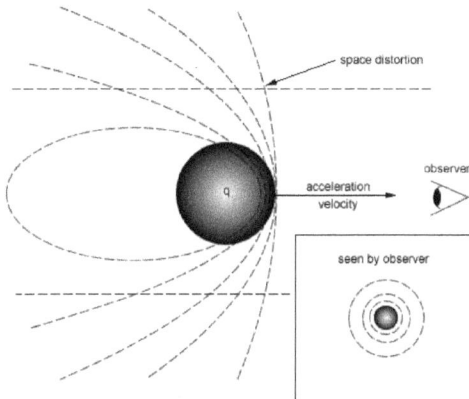

Figure 4: Magnetic field is a medium distortion field arising from distortion due to movement in the medium

$$D_t = \frac{\Delta t_{back}}{\Delta t_{front}}$$

(49)

Neutrons in an atomic system orbit the core nuclei protons faster than c, while electrons orbit slower than c. This is reasoned to have a significant effect, due to the Biot-Savart Law, on the field strength and collective cross-sectional area of the nucleus. The Earth magnetosphere's compressed radius is between 6.5×10^8 [m] to 9.0×10^8 [m] which is comparable to the radius of the Sun. Earth also has a satellite, the moon, which is reasoned to increase its magnetosphere size, meaning planets with no satellites have smaller magnetospheres. This dynamic would also play been a star and its satellites. The more satellites a star (cosmological proton(s)) has, or the more massive, the larger and stronger its magnetosphere (heliosphere) is.

The cross-sectional area of this distortion field summed with the cross-sectional area of the proton in a Deuterium system collectively drag the system and produce inertial mass for the system close to *double* the inertial mass of a proton alone. The cross-sectional area of the neutron distortion field is roughly the cross-sectional area of the proton.

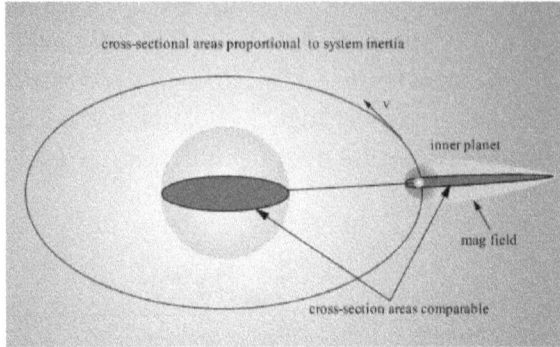

Figure 5: Combined cross-sectional areas of cosmological proton (star) and magnetosphere of rock planet (neutron) impede ISM (CM) particles and drag system through ISM (CM). Similar dynamic occurs are the quantum scale with the quantum medium (QM) and is cause for system inertia.

The summed cross-sectional area of the system is the area that imposes resistance against the ISM (cosmic medium) or quantum medium respectivey. This means the capability of ISM particles to penetrate through these magnetosphere (including the magnetotail) is less than than in vacuum space. This dynamic translates at the quantum scale to motion resistance against the quantum medium and results in inertial mass.

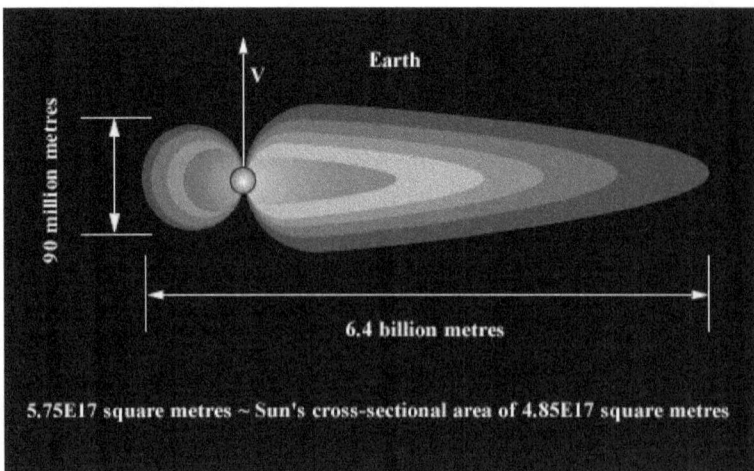

Figure 6: Cross-sectional area of inner system magnetospheres, including tails, are comparable to Sun's cross-sectional area. Inner

The magnitude of Earth's (neutron's) magnetosphere (distortion-sphere) is 14 times weaker than Jupiter's (electron's). Jupiter's magnetosphere covers a much larger area than the Sun's cross-sectional area, which at the quantum scale adds to the collective inertial mass of the atomic system. This is why some atoms have higher inertial mass than the sum of their constituent quantum particles alone.

It is important to note that Earth's Moon has an orbital radius of 3.8447×10^8 [m] giving it an area of 4.6438×10^{17} [m²] which is very close to the cross-sectional area of the Sun. This eludes to the connection that the distortion cross-sectional area is derived from the gravitational potential well of the planet, and that the planet's magnetosphere field is an artefact (a *membrane*) of the distorted quantum medium. Distortion is ultimately the cause of gravity as it produces a density-pressure gradient the medium.

Due to the fact that inner system objects orbit much more rapidly than outer system objects, they cover more area in a given period of time making the inner system more resistant to ISM (cosmic medium) penetration. Again the same dynamic occurs at the quantum scale where the nucleus is less penetrated by the quantum medium than the outer system of electrons.

Considering the following ratio relation using an unbound Jupiter class gas-planet moving in free interstellar space, or cosmic medium (*CM*), and its quantum counterpart (electron) moving in the quantum medium (*QM*) both at velocity v and acceleration a using equation (34) for inertial mass of quantum particle in motion:

$$\frac{m_e}{q_e} \approx \frac{m_{J\,(CM\,drag)}}{m_J}$$

$$\frac{\frac{1}{2}C_{d_{QM}}\rho_{QM}\left(at^2\right)\left(\pi r_e^{\,2}\right)}{q_e} \approx \frac{\frac{1}{2}C_{d_{CM}}\rho_{CM}\left(at^2\right)\left(\pi r_J^{\,2}\right)}{m_J} \qquad (50)$$

Or by using the relative *equivalence principle* in the form of pressure P from the quantum medium and cosmic medium (ex. ISM) respectively:

$$\frac{m_e}{q_e} \approx \frac{m_J (CM \; pressure)}{m_J}$$

$$\frac{\left(\frac{P_{QM}}{a}\right)(\pi r_e^2)}{q_e} \approx \frac{\left(\frac{P_{CM}}{a}\right)(\pi r_J^2)}{m_J} \tag{51}$$

Where equation (50) or (51) reduces to:

$$u_e\left(\frac{r_e^2}{q_e}\right) \approx u_J\left(\frac{r_J^2}{m_J}\right) \tag{52}$$

Which remarkably results in the following:

$$m_e = u_e r_e^2 \approx u_J\left(\frac{r_J^2}{m_J}\right) q_e = u_J (4.1256x10^{-31}) \quad [kg=C] \tag{53}$$

The value of **4.125x10⁻³¹** is very close to **9.1x10⁻³¹**. Given the rest of this frame-work, *this is beyond coincidence*. To equal an electron mass, u_J would have to be:

$$u_J = \frac{9.1x10^{-31}}{4.125x10^{-31}} = 2.20574 \quad [(kg=C).m^{-2}] \tag{54}$$

Can it be merely defined that **inertial mass to charge ratio of a quantum particle is the ratio between the cross-sectional area, or specifically the radius squared, of the quantum particle to its charge (quantum mass)**? It would appear that the ratio between the radius squared over mass (or charge if scaled by S) is roughly a consistent quantity amoungst all cosmological objects, both a the macroscopic and quantum scale. This value is 5.6797x10⁻¹².

Here are some results solely using this logic:

object	(Radius^2 / Mass) x e kg	Similarity to e %	Grouped %
Mercury	2.89E-030	31.49%	31.49%
Venus	1.21E-030	75.21%	75.21%
Earth	1.09E-030	83.49%	
Moon	6.59E-030	13.81%	97.30%
Mars	2.88E-030	31.60%	31.60%
Jupiter	4.30E-031	47.25%	
Io	6.22E-030	14.63%	
Europa	8.21E-030	11.09%	
Ganymede	7.50E-030	12.13%	
Callisto	8.65E-030	10.52%	95.62%
Saturn	1.03E-030	88.35%	
Titan	7.90E-030	11.51%	99.86%
Uranus	1.27E-030	71.65%	
Oberon	3.08E-029	2.95%	
Titania	2.82E-029	3.22%	
Umbriel	4.67E-029	1.95%	
Ariel	4.23E-029	2.15%	
Miranda	1.41E-028	0.65%	
Puck	3.28E-028	0.28%	82.85%
Neptune	9.96E-031	91.37%	91.37%

Table 4: Radius squared on mass ratio multiplied by elementary charge gives the quantum particles inertial mass.

The resulting value of all gas planets in table (4) are remarkably close to the value of an electron mass 9.1x10^{-31} [kg]. Even rock planets are remarkably close, which eludes that this formulation is not yet complete, or to something unexpected. As an *initial rule of thumb*, this relation between a quantum particle's mass and charge can be used in the meantime.

It is also notable that the sum of the similarity percentages for Mercury and Venus equals 106.7%, which eludes the possibility that Mercury was the moon of Venus at some point in the past.

In attempt to explain why a neutron (rock planet) has larger inertial mass than an electron even though a gas-giant has a larger cross-sectional area than a rock planet, is to take into account the pressure variable P in equation (51), acceleration variable a, or the combination. The pressure variable, or acceleration variable, must be different in the regions around a gas-giant and rock planet. If acceleration remains constant, pressure must be larger around a rock planet which may also account for high magnitude magnetic fields.

$$m_e = \frac{\left(\dfrac{P_{outer\ system}}{a}\right)(\pi\, r^2_{gas})}{m_{gas}} < m_n = \frac{\left(\dfrac{P_{inner\ system}}{a}\right)(\pi\, r^2_{rock})}{m_{rock}} \tag{55}$$

Specifically the pressure/acceleration (P/a) ratio between the inner and out system would be valued at around 1840.66. It is reasoned that the mass-density ρ and orbital velocity v of the object play into this factor as:

$$P \propto \rho\, v \tag{56}$$

Taking equation (48) on distortion, we get the following in determining a radius of distortion:

$$D = \frac{u_D}{3}\left(\frac{M}{\frac{4}{3}\pi\, r^3}\right)|v||r|\sin(90^\circ) = \left(\frac{u_D}{4\pi}\right)\left(\frac{M\,v}{r^2}\right) \tag{57}$$

$$D = \left(\frac{u_D}{3}\right)\rho\, v\, r$$

The addition to the radius of inner system objects (rock planets/neutrons) making it inline with neutron inertial mass (1840 larger than electron) appears to be related to density and velocity of the these inner objects. The denser the object, the less the quantum medium can penetrate it, therefore the larger the distortion. The average density for main outer gas planets is 1134.57 [kg/m³], for main inner rock planets it's 5015 [kg/m³].

$$P \approx \rho\, v = \frac{3\,D}{u_D\, r} \tag{58}$$

It is important to note that the inner system objects travel faster the surrounding medium's wave velocity c, for cosmic medium and quantum medium respectively. Only when $v > c$, a *shock wave* forms in the medium creating an area of lower medium pressure density in the tail surrounded by a membrane of higher pressure density by an angle of:

$$\theta = \arcsin\left(\frac{c}{v}\right) \tag{59}$$

Since the shock-wave is formed only when $v > c$, then inner system objects distort the medium to a greater degree than outer system objects, especially if they were to become unbound from the system at their high velocities.

Now we consider the scaling framework transformations, equations (23) (25) (43), on the radius mass and density of unbound planetary object. Here the requirement is to meet a radius that will give proper ratios to meet the inertial mass of neutrons, protons and electrons using the Solar System planets and elementary charge as done in equation (53):

object	Radius Required m	Velocity Required m/s	Velocity Required c	Resulting Mass kg	Resulting Density kg/m^3
Mercury	5.8760E+07	14858.15	0.8581	7.94E+024	9.35
Venus	2.2558E+08	9252.67	0.5344	1.81E+026	3.77
Earth	2.4987E+08	7725.93	0.4462	2.34E+026	3.58
Moon	2.7716E+07	10758.57	0.6214	1.17E+024	13.14
Mars	8.1888E+07	7477.83	0.4319	1.55E+025	6.73
Jupiter	1.0383E+08	15817.94	0.9136	2.76E+027	588.72
Saturn	5.6814E+07	14111.91	0.8150	5.35E+026	696.71
Uranus	2.2205E+07	7733.32	0.4466	7.36E+025	1604.23
Neptune	2.4118E+07	6165.34	0.3561	9.79E+025	1666.29

Table 5.: Velocity required using scaling framework transforms by system objects to meet ratio required to meet inertial mass of neutron and electron respectively ($1.675x10^{-27}$ [kg], $9.1x10^{-31}$ [kg]).

As can easily be seen from Table 5, the resulting mass-density of the inner objects, whose resulting velocities fall under c_s (17314.56 [m/s]) cosmic medium (ISM/ IGM) wave velocity (scale relative to speed of light c in the quantum medium), becomes extremely low compared to the outer system objects. *An interesting mass-density pattern definitely emerges.* Essentially such a drastic change in density can be considered a form of trans-mutating decay.

Now we consider dimensional ramifications. If the electron inertial mass is indeed based *only* on the cross-sectional area of the electron, or the square product of its radius, and its subsequent sub-quantum medium distortion field (quantum scaled magnetic field), rendering $u_J = 1$ and dimensionless in equation (53), then the dimensionality of mass has gotten wonderfully more perplexing where $L^2 = M$ (length squared = mass). Using this rational, the macroscopic radius of an unbound moving electron using S only is:

$$m_e \rightarrow r_e^2$$
$$r_e \rightarrow \sqrt{m_e} = 9.5394 \text{x} 10^{-16} [\sqrt{kg} = m]$$
$$r_e' = r_e S = 1.0517 \text{x} 10^8 [m = \sqrt{kg}]$$
(60)

Macroscopic mass of electron using S only:

$$m_e' = q_e S^2 = 1.9475 \text{x} 10^{27} [kg = C]$$
(61)

Giving it a macroscopic mass density ρ_e' in interstellar space of:

$$\rho_e' = \frac{m_e'}{\frac{4}{3} \pi (r_e')^3} = 399.6808 [(kg = C)/m^3]$$
(62)

It would appear that the dimensionality of mass and charge is further based on L^2 (length squared) and not solely M (mass) or Q (charge), where ratios, if only certain ratios, appear to be invariant between discrete scales of S.

If equation (54) is not constant, or valued 1, then it must be variable. If this is the case, it is reasoned to be dependent on the quantum medium density of the surrounding space, where at Jupiter u is equal to 2.20574, but in the inner system, u has a different value.

8.8 Gravitation Mass

Gravitational mass is similar to inertial mass. This is described by the Equivalence Principle. Inertial mass is experienced by an object accelerating while gravitational mass is experienced by pressure on an object at rest, possibly on another object. The quantum medium exerts a pressure on the mass object resting on the surface of an other object. This pressure is related to the cross-sectional area of the atomic nuclei and quantum particles in the object.

$$P = \frac{F}{A} \quad \rightarrow \quad F = P A \quad \rightarrow \quad m_g = P \left(\frac{A}{a} \right)$$
$$m_g \propto m_i = \left(\frac{C_d \rho_{qm} v^2}{2} \right) \left(\frac{A}{a} \right)$$
(63)

Where F is the force exerting pressure, P is pressure exerted by the quantum medium on a quantum particle, and A is the cross-sectional area of the particle (or product of radius squared).

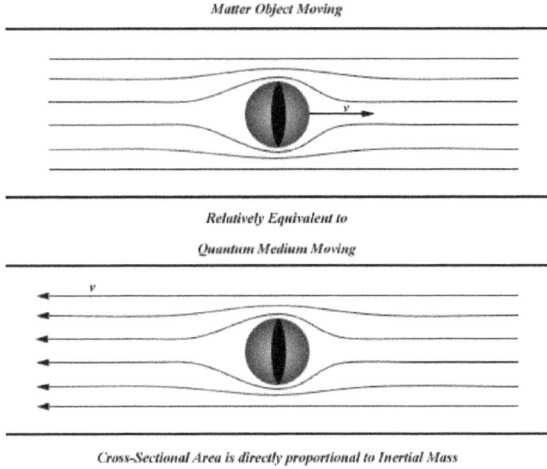

Figure 7: Equivalency Principle dependent on cross-sectional area of quantum particles, or sum of cross-sectional area for macroscopic inertia.

Gravitational mass is derived by the uniform pressure of the surrounding quantum medium on the sum of cross-sectional area of quantum systems in macroscopic matter objects.

8.9 Momentum

If quantum medium (QM) opposes motion of matter in the form of drag, then how can constant momentum exist (Newton's First Law)? It is highly reasoned that macroscopic matter motion distorts not only the quantum medium (QM) but also the collective quantum medium (CQM). The moving object produces a distortion field in smaller discretely scaled spatial material mediums by factors of S, like the sub-quantum medium (SQM), which then shields the object from the quantum medium's sub-quantum particles by "slipping" them around the object along the distortion field lines, just like a magnetic field does with charged particles from the ISM. This effect, and constant momentum, is broken only when acceleration is applied, and when an object's velocity approaches the wave velocity of the spatial material medium which alters (increases) the drag coefficient (C_d) of the medium on the object. Inertial mass is resistance to acceleration, not constant velocity as equation (34) predicts.

8.10 Passage of Time

To derive a passage of time difference between the two self-similar scales, we take into account the orbital velocity of the outer most valence electron (q) in a Beryllium atom in ratio to the orbital velocity of the outer most planet (Neptune) in the Solar System and the average orbital velocity of the Asteroid Belt (c_s = 17314.5158 [m/s]) multiplied by the speed of light c:

$$v_q = \left(\frac{v_{(planet\ scale\ relative)}}{c_s} \right) c = (5430[m/s])(17314.5158)$$

$$v_q = (0.26) c [m/s]$$

(64)

Neptune has an orbital period of 60,190.03 [days], which is 5.2004x10⁹ [seconds/orbit]. The orbital period for the outer most electron in a Beryllium atom is the time taken to travel the circumference of the atom:

$$C = 2\pi r = \pi \left(\frac{r_{Neptune}}{S} \right) = 2\pi \left(\frac{4.5539 \times 10^{12}[m]}{S} \right)$$

$$C = 2.5953 \times 10^{-10}[m]$$

(65)

Using the velocity in equation (64) with the circumference in equation (65) we get the following orbital period for a valence electron:

$$t = \frac{d}{v} = \frac{2.5953 \times 10^{-10}[m]}{(0.3136)c[m/s]} = 2.76052 \times 10^{-18}[s]$$

(66)

This gives a passage of time difference between the two scales of:

$$T = \frac{5.2004 \times 10^{9}[s]}{2.76052 \times 10^{-18}[s]} = 1.8838 \times 10^{27}$$

(67)

The same set of calculations, from equations (64) to (67), can be applied to Pluto the result is very close to (67). The passage of time difference value in equation (67) is also closely related to the speed of light c and π:

$$\frac{\log(1.8838 \times 10^{27})}{\log(299792458)} = 3.2176 \approx 3.14159... = \pi$$

$$(97.64 \ percent)$$

(68)

Using Pluto and the atomic radius of Beryllium of 105 [pm] as the reference, the value (3.18) is even closer to π. If π is the correct value, than this small deviation in value speaks to the possibility that the outer planets in the Solar System are in one particular state of a possible many of which the average is π.

Therefore like equation (3), an equation for the passage of time is derived as:

$$T = T = \left(\frac{c}{1[m/s]}\right)^{\pi} = 4.2728 \times 10^{26}$$

(69)

Of which the relative velocity equation is formulated like equation (5):

$$\tau(v) = \left(\frac{v}{v_o}\right)^{\pi} = (\gamma)^{\pi}$$

(70)

Where 1 macroscopic second would account for T quantum seconds passed at the quantum scale relatively.

9   Uncertainty Principle

The passage of time factor T is so large that predicting the position and moment of electrons compared to gas-giants is extremely difficult. This accelerated passage of time at the quantum scale significantly contributes to quantum particles behaving as waves as well as particles. The accelerated passage of time is so extreme for a quantum particle that the relative passage of time for a cosmological scale relative is 1.35×10^{19} years for every relative macroscopic second. This immense passage of time for cosmological planets would encompass the entire spans of geological changes including potential de-constitution and reconstitution of the whole object having a wave-like characteristic in its overall physical evolution overtime. From a macroscopic scale perspective, quantum particles would have a constantly changing physical structure due to this time difference. This aids in explaining the duality of quantum particles and the fundamental cause why contemporary quantum probability theory is inherently used with the Uncertainty Principle. Future predictions can be more deterministically tooled with this new framework by slowing down quantum dynamics mathematically.

For minimum range of error (deviation), or the uncertainty, for position and momentum is:

$$\sigma_x \sigma_p \geq \frac{\hbar}{2} \tag{71}$$

For minimum range of error, or the uncertainty, for time and energy is:

$$\sigma_E \sigma_t \geq \frac{\hbar}{2} \tag{72}$$

Where $2/\hbar$ is a very large number (1.8965×10^{34} s m^{-2} kg^{-1}). What this means is change in position of a distant planet, or free-roaming planet, is less known the greater the change in momentum. Also, the change in energy of a planet is less known the larger the change in time. By combining the two equations (71) and (72), we get the following:

$$\sigma_E \sigma_t \propto \sigma_x \sigma_p \tag{73}$$

The deviation in the certainty of time is proportional to the actual passage of time. Simplest way to understand this logic is by attempting to predict time without reference to some absolute precision clock. The more actual time that passes as observed by a third party with an absolute precision clock, the more deviation in the certainty of time (erroneous) introduces itself for someone without absolute precision time keeping instrumentation. Another way to understand this is to classify certainty as deterministic and uncertainty as non-deterministic.

$$\sigma_t \propto \Delta T \tag{74}$$

Where ΔT here is passage of time and σt is the deviation in the certainty of the time. This makes equation (74) the following (with σ standing for deviation):

$$\sigma_E \Delta T \propto \sigma_x \sigma_p \tag{75}$$

Where it can be restructured as:

$$\sigma_x \propto \frac{\sigma_E}{\sigma_p} \Delta T \tag{76}$$

Where it can be further reduced to:

$$\sigma_x \propto \frac{\sigma_v}{2} \Delta T \tag{77}$$

Equation (77) means that as the passage of time increases, the *uncertainty* in position of a planet increases if *uncertainty* in velocity remains constant. Also, the *certainty* in velocity increases as the passage of time increases if the *uncertainty* in position remains constant. This is a self-referencing equation as one understands that position divided by time is velocity, therefore making this equation chaotic which is the basis for quantum theory.

10 Charge-Mass Relation

Electric force is produced by the density-pressure potential gradient of the sub-quantum medium (*SQM*), and gravitational force is a fractal scale relative force to electric force arising from the density-pressure potential gradient of the quantum medium (*QM*). The quantum medium cannot penetrate quantum particles, but the sub-quantum medium can.

Inertial mass, and the mass-charge relation, is best approached using a charged quantum particle where an electric field is applied to a charged quantum particle until the particle researches a terminal velocity due to quantum medium dragging force resulting in a zero net force acting on the particle. This determines the parameters inertial mass is dependent on.

$$F_E - F_D = F_{net} = 0 \;\rightarrow\; qE = \frac{1}{2} C_d \rho_{qm} v^2 A = m a_A \tag{78}$$

Where a_A is the measurable apparent acceleration of the particle, v is the measurable apparent velocity in the same direction as a_A, C_d is the drag coefficient of the quantum medium, q is charge of the particle, E is the electric field, ρ_{qm} is the mass-density of the quantum medium, A is the cross-sectional area of the quantum particle (cross-sectional charge area), and m is the perceived mass of the quantum particle. Inertial mass can be defined:

$$m = \frac{\left(\frac{1}{2} C_d \rho_{qm} v^2 A \right)}{a_A} \tag{79}$$

Where

$$v = at \;\rightarrow\; v_2 = (at)^2 = a^2 t^2 \tag{80}$$

Combining (79) and (80):

$$m = \frac{\left(C_d \rho_{qm} A (a_A^2 t^2) \right)}{2 a_A} = \frac{1}{2} C_d \rho_{qm} A (a_A t^2) \tag{81}$$

Equation (81) shows that inertial mass m is only measurable under acceleration a_A and proportional to the cross-sectional area A of the quantum particle. This permits for momentum at velocity v while acceleration is zero. Combining equation (23) with (81) to the cross-section radius:

$$m = \frac{1}{2} C_d \rho_{qm} (a_A t^2) \left(\pi \left(\frac{r_o}{s(v)} \right)^2 \right) = \frac{\frac{1}{2} C_d \rho_{qm} (a_A t^2)(\pi r_o^2)}{s(v)^2} \tag{82}$$

Quantum particle inertial mass is a product of the particles cross-sectional area (radius squared), velocity and the charge-density of the quantum medium. The quantum particle's inertial mass is derived differently than that of macroscopic inertial mass (drag), but this framework indicates that the dimension of mass (M; kilogram) applies to both the macroscopic mass and the contemporary dimension of charge [Q]. **The dimension of mass (M) is a *measure of resistance to motion under acceleration* irrespective of the medium imposing resistance; quantum or sub-quantum mediums**. This requires a redefinition of the dimensions of mass (M) and charge (Q) as both are the same dimension. Charge becomes quantum-mass which is equal dimensionally to macroscopic-mass. The emphasis is now placed on the means of applied force; electric or gravitational potential. Therefore equation (78) can be rewritten as such:

$$m_q = q(1[kg = C]), \quad a_E = E(1[C = kg])$$
$$m_q a_E = m a_A \tag{83}$$

The same equations (81 to 82) can be applied to particle charge in the sub-quantum medium where ρ_{sqm} is charge density of the sub-quantum medium.

11 Special Relativity

Equation (82) also supports the Lorentz factor and Einstein's mass transform if the drag coefficient C_d is explored more thoroughly where the wave velocity of the medium is c. The Lorentz factor is proportional to the drag coefficient graph up to Mach 1.0 in Figure 8.

Where the Lorentz factor can be recreated using drag force. At velocity c is when drag force is maximum, therefore in ratio:

$$\frac{F_{D_{cv}}}{F_{D_c}} = \frac{F_{D_c} - F_{D_v}}{F_{D_c}} = \frac{v_{cv}^2}{c^2} = \frac{c^2 - v^2}{c^2} \quad \rightarrow \quad \frac{v_{cv}}{c} = \sqrt{1 - \frac{v^2}{c^2}} = \gamma \tag{84}$$

Applying equation (84) to applied force a mass object gradually approaching velocity c where contemporary theory predicts that no amount of applied force can make any mass object travel faster than c:

$$F_{net} = F_{applied} - F_{drag} \quad \rightarrow \quad F_{net} = \gamma \left(F_{applied} - F_{drag} \right) \tag{85}$$

Manipulating equation (85) with the understanding that drag force was not initially considered, or has since been omitted, resulting in applied force being the only considered force. Therefore net force was equal to applied force:

$$\frac{F_{net_o}}{\gamma} = F_{applied} - \left(F_{drag \rightarrow 0} \right) = F_{net} \tag{86}$$

If applied acceleration is constant, yet movement slows as the object approaches c, then the following mass transform equation (87) is devised from equation (86), and is consistent with Special Relativity:

$$m = \frac{m_o}{\gamma} = \frac{m_o}{\sqrt{1 - \frac{v^2}{c^2}}} \tag{87}$$

The problem with equations (84), (85), (86) and (87), and Special Relativity, exists at and after Mack 1.0. Figure 8 indicates that the increased inertial mass due to increased drag coefficient from approaching the quantum medium wave velocity

c, and drops off after Mach 1.0. The Lorentz factor becomes imaginary after Mach 1.0 which is difficult to tangibly translate to into physical reality. Another problem with Special Relativity is the lack of data and experimentation involving a macroscopic matter object, not quantum object, reaching the light speed and the gradual affects experienced on the object as it approaches c.

Special Relativity, and general relativity, is the fluid dynamic treatment of matter object movement in the quantum medium, like a rock moving through air. The faster the velocity, the higher the air pressure on the surface of the rock.

$$p \propto v \tag{88}$$

The quantum medium acts in a similar fashion to air, but with the difference that it deeply penetrates macroscopic matter yet cannot penetrate fermionic matter constituting the macroscopic matter. Due to this scale nature of the quantum medium, the affect on a moving macroscopic matter object is different than the affect on a quantum matter object. Increasing the velocity of a macroscopic object through the quantum medium will increase the external pressure on the bonded constituent fermion particles pushing them closer together and scaling the overall macroscopic object smaller. This external pressure is described in following equation (99). The quantum medium is fundamentally more complex than air and classical definitions of luminiferous ether. Density changes in the quantum medium affect the experience of space (distance) and time. Further more, Special Relativity diverges from fluid dynamics with the open assumption that fluid wave velocity, light speed in this case, is an absolute limit.

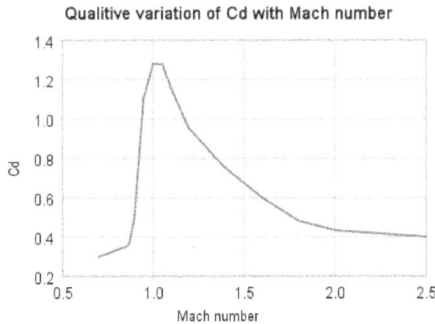

Qualitive variation of Cd with Mach number

Figure 8: Drag coefficient vs. velocity. The drag coefficient increases as it approaches Mach 1, or speed of sound. This is analogous to a similar effect in the quantum medium as an object approaches the speed of light.

47

The effect on drag coefficient C_d is somewhat defined by the following equation where M here is Mach velocity, and I is the modifying intensity factor of C_d:

$$I=\frac{1}{\sqrt{1-M^2}}, \quad M<1$$

$$I=\frac{1}{M-1}, \quad M>1 \tag{89}$$

$$M=mach=\frac{v}{c}$$

Figure 8 is eerily similar to the graph of black-body radiation at high temperature (5000 K):

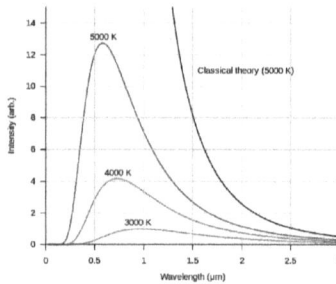

Figure 9: High temperature radiation has a similar graphical pattern to drag coefficient vs. velocity, which is similar to Lorentz factor in special relativity and its effect on inertial mass.

This may elude to better derivation of a transformation function in replacement of the Lorentz factor or equation (89). Planck's black-body radiation intensity (I) equation is:

$$I(f,T)=\frac{2hf^3}{c^2}\left(\frac{1}{e^{\frac{hf}{kT}}-1}\right) \tag{90}$$

Where $I(f,T)$ is the energy per unit time (or the power) radiated per unit area of emitting surface in the normal direction per unit solid angle per unit frequency by a black body at temperature T, h is the Planck constant (6.626068×10^{-34} [m^2 kg/s]), c is the speed of light in a vacuum, k is the Boltzmann constant ($1.3806503 \times 10^{-23}$ [m^2 kg s^{-2} K^{-1}] or [J/K]) , v is the frequency of the electromagnetic radiation, and T is the temperature of the body in kelvins.

Using formulation (90) as a reference, the following equation is approximately derived, with velocity in units of c, as a dimensionless modifier to the drag coefficient (C_d) found in the drag force equation:

$$I(v) = \frac{2.1}{\left(\dfrac{v - 0.645[c]}{1[c]}\right)^3 \left(e^{\left(\frac{1[c]}{v - 0.645[c]}\right)} - 1\right)} \tag{91}$$

Equation (91) results in the following graph which is similar to the drag coefficient vs. velocity graph in Figure 8.

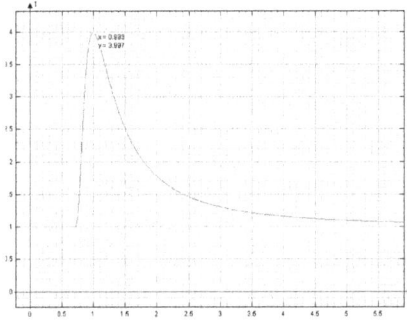

Figure 10: Inertial mass drag coefficient modifier in velocity units of [c]

This equation (91) has values to fit the magnitude of change represented in Figure 8, but it can be generalized with variables to represent these numerical values $(\alpha=2.1, \beta=0.645)$:

$$I(v) = \frac{\alpha}{\left(\dfrac{v - \beta[c]}{1[c]}\right)^3 \left(e^{\left(\frac{1[c]}{v - \beta[c]}\right)} - 1\right)} \tag{92}$$

Where drag force equation (33) would be modified to include equation (91):

$$F_d = \frac{1}{2}\left(IC_d\right)\rho_{qm} v^2 A = m a_d \tag{93}$$

Which would result in a change in the inertial mass equation (34) derived from drag force through the quantum medium:

$$m = \frac{(I C_d)\rho_{qm} v^2 A}{2a} = \frac{1}{2}(I C_d)\rho_{qm}(a t^2) A \qquad (94)$$

11.1 Co-Existing with Einstein

As has been demonstrated, the mechanistic theoretical basis for the effects of special relativity, specifically those of mass increase and time dilation, can be explained alternatively in relation to fluid dynamical effects on the medium's drag coefficient. This alternative explanation can coexist with the velocity transform equations of this framework, because the transforms of this framework are dependent on the natural terminal velocities in space, as described by equations (8) to (22). These terminal velocities are the result of quantum materialistic interaction with a medium of sub-quantum particles, called the quantum medium (QM), and the centrifugal forces (solar wind) that push macroscopic matter objects naturally into motion in cosmological systems. By changing the density of this medium, the terminal velocities permitted in those spaces are also changed affecting the macroscopic matter object's size and experience of the passage of time. Equation (95) demonstrates special relativity modifier (*I*) and this framework's scaling transform (*s(v)*) working in conjuncture. Given this understanding, both the formulations from Einstein's successful theories and this framework can co-exist.

$$m = \frac{\frac{1}{2}(I C_d)\rho_{qm}(a t^2)(\pi r^2)}{s(v)^2} \qquad (95)$$

12 Mechanistic Scaling

A moving fermion quantum particle distorts the collective quantum medium (*CQM*) it moves through, as a rock does moving through a dust cloud, air or water. The distortion of solely the quantum medium (*QM*) is more apparent than the collective sum of all discretely scaled spatial mediums. This manifests the effect classified as a magnetic field. This field is caused by the distortion of the multiple fractal levels of self-similar mediums forming the collective quantum medium. The distortion of one level contributes to the distortion of the subsequent level forming a stable field. This distortion at the macroscopic level, the magnetic field, acts as insulator (shielding effect) from free unbound charged particles outside the system, and traps charged particles inside the system. Strong magnetic fields also affect neutral particle systems (atoms). This dynamic subsequently forms a localized dragging effect of the quantum medium inside the field governed by the inverse square law; more classically known as inertial frame dragging.

The scaling of a macroscopic object is dependent of the strength of the distortion field cause by its quantum constituents. The objects travelling in the same direction each form a distortion well in the quantum medium where the quantum medium between these particles is less dense than the surrounding quantum medium causing a attractive external pressure on the particles reducing their distance apart. This field strength is dependent on the velocity of the particles which is partially governed by the Biot-Savart Law.

$$\vec{B} = \left(\frac{\mu_o}{4\pi} \right) \frac{q}{r^3} (\vec{v} \times \vec{r})$$

(96)

Where

$$F = q\vec{v} \times \vec{B}$$

(97)

Where for two quantum particles travelling parallel to each other a v:

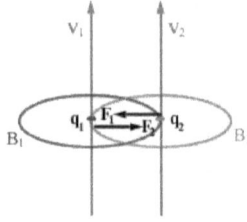

Figure 11: Mutual magnetic forces on parallel moving charged particles

Combining (96) and (97)

$$F = q\,\vec{v} \times \left(\left(\frac{\mu_o}{4\pi} \right) \frac{q}{r^3} (\vec{v} \times \vec{r}) \right)$$

(98)

Where vectors are a right angles at double the magnitude for 2 objects

$$F = (2)\, q_1 |v_1| \left\| \left(\left(\frac{\mu_o}{4\pi} \right) \frac{q_2}{r^3} (|v_2||r| \sin(90^o)) \right) \right\| \sin(90^o)$$

$$v = v_1 = v_2$$

$$F = \left(\frac{\mu_o v^2}{2\pi} \right) \frac{q_1 q_2}{r^2}$$

(99)

Equation (99) is related to equation (57), and looks eerily similar to Coulomb's electric force equation.

The symmetry of scaling is dependent on a few factors. The velocity of the object, the mechanism of propulsion, and the magnitude of fractal distortion. It is reasoned that self-propelled objects will scale more effectively than externally propelled object (ex. via induced electric field). If only the quantum medium is considered and subsequently distorted, the scaling is asymmetric and Einstein relativistic dynamics play a more dominate role. If the collective quantum medium is distorted, the scaling is symmetric where this scaling framework plays a greater role. The additional effect from a distortion field is due to shielding; the absorption and release of quantum medium constituents (sub-quantum particles) onto the quantum particles. The stronger the field, the less quantum medium matter the particles absorb while still releasing quantum matter through *energy*.

13 Transmutation

Table 3 indicates that matter objects can trans-mutate into a another type of matter object (larger or smaller) simply through a change in kinetic velocity, which is a terminal velocity in the quantum medium derived from a change in quantum medium density. This gives an explanation for quantum particle transmutation (particle decay), as well as supporting *Expanding Earth theory*.

The Feynman diagram for beta decay visually details this decay of $n^0 \rightarrow p^+ + e^- + \nu^e$.

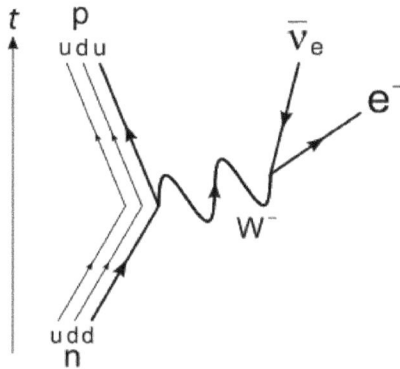

Figure 12: Feynman Beta Decay Diagram

Figure 12 describes the beta-decay process in weak force interaction of a neutron (*n*) decelerating, due to interaction with denser quantum medium (possibly a boson), and trans-mutating itself (or decaying depending on the perspective) through absorption of the quantum medium sub-quantum particles into one of the W boson types. The W boson is understood to be short lived due to it being a scale relative to a nebula in an accelerated time-frame of reference (*T*). This newly formed W boson (nebula) expels, or constitutes, an electron (*e*, gas-giant) and electron neutrino (ν^e, gas-giant moon) from its own material, while its remaining bulk matter accelerates again trans-mutating (constituting) into a proton (*p*). Thus, neutron decay produces a proton, electron and an anti-neutrino. Re-interpreting figure 12 is:

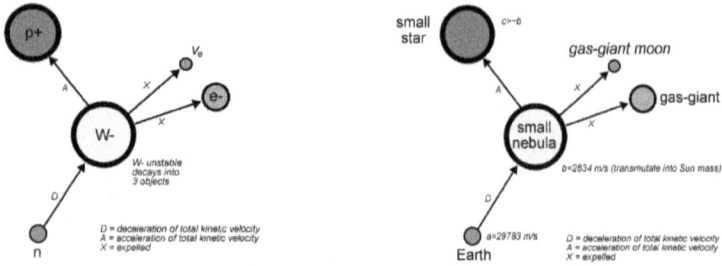

Figure 13: Re-interpreted Feynman beta decay diagram quantum scale

For example, at the macroscopic scale, if the Earth (neutron) were to experience an increase in quantum medium density, effectively slowing its natural velocity through space from 29,783 [m/s] to 2834 [m/s], the Earth's mass would increase to that of the Sun's (using mass transform equation), with a resulting radius larger than the Sun's. In this expanded state, the resulting Earth would resemble a small nebula due to its low mass-density. Given sometime, this nebula would eventually condense into smaller planetary objects. It can be surmised from quantum beta decay that these macroscopic objects produced from the nebula are a cosmological proton, gas-giant and gas-giant moon.

14 Force

Classical force equation is simply:

$$F = ma = m\left(\frac{\Delta v}{\Delta t}\right)$$

$$(100)$$

Dimensional analysis reveals force expressed as such:

$$F = MLT^{-2} = M\left(\frac{L}{T^2}\right)$$

$$(101)$$

Force has two time components (T^2) described in its acceleration variable defined by the unit of seconds squared. There is also a mass component (M) and distance component (L).

As an object accelerates towards the speed of light, by experiencing a significant change in quantum medium density, its mass decreases, its collective size decreases and its perceived passage of time increases from its initial reference frame; initial quantum medium density. The distance component (L), the measurement of, in relation to the object's velocity and acceleration, is invariant at all scales thus has no transformation as velocity increases. This logic derives equation (102) by combining mass transform equation (36) and time transform equation (70).

$$F = \left(\frac{M}{s(v)^2}\right)\left(\frac{L}{1}\right)\left(\frac{t(v)}{T}\right)^2$$

$$(102)$$

Therefore equation (102) describes in an increase in relative force because the communicators of force, the causal dynamics of, at the quantum scale travel faster than at the macroscopic scale. The factor by which time changes is greater than the factor by which scale changes.

$$F = mg\left(\frac{\tau(v)^2}{s(v)^2}\right) \rightarrow F = qE \quad , v_{total} \approx c$$

$$(103)$$

In this framework, mass and charge are the same "thing" (dimension), describing a spatial medium resistance to motion, at two different points on the velocity spectrum (v and v'), therefore in equation (104) mass (Mm), charge (q_1q_2) and distance (d) can be eliminated leaving only the transforms and constants. This analysis takes two identical mass pairs. One pair remains stationary in its perceived macroscopic mass while the other pair is moving collectively near c in its quantum charge state. If they were to exist at the same velocity, they would have the same mass.

$$\frac{(\delta G)\,Mm}{d^2}\left(\frac{t(v')^2}{s(v')^2}\right)=\frac{k_e q_1 q_2}{d^2} \tag{104}$$

Where taken $\delta=1$, meaning the density and mass of M and m are equal.

$$\frac{k_e}{G}=\frac{t(v')^2}{s(v')^2} \tag{105}$$

Given equation (41), the left and right of equation is dimensionless.

$$\frac{k_e}{G}=-\left(\frac{v'}{v}\right)^{(2\pi-2e)}=\left(\frac{v'}{v}\right)^{(0.8466)} \tag{106}$$

This equates to:

$$\frac{v'}{v}=\left(\frac{k_e}{G}\right)^{\left(\frac{1}{0.8466}\right)}=\left(\frac{k_e}{G}\right)^{(1.1812)}$$

$$=\left(\frac{8.9876\mathrm{x}\,?\,10^9\ [N.m^2/C^2]}{6.673\mathrm{x}\,?\,10^{(-11)}\ [m^3/kg/s^2]}\right)^{1.1812}$$

$$=\left(\frac{8.9876\mathrm{x}\,?\,10^9\ [m^3/kg/s^2]}{6.673\mathrm{x}\,?\,10^{(-11)}\ [m^3/kg/s^2]}\right)^{1.1812} \tag{107}$$

$$=5.9810\mathrm{x}10^{23}$$

It is very interesting to note that the resulting value in (107) of **5.9810x10²³ is extremely close to Avogadro's constant** N_A of 6.02214x10²³ at 99.32% similarity. It appears that the proportionality on the differences of dimension between gravity and electric force result in what appears to be a completely unrelated constant describing the number of base constituents in a mole of matter.

$$v' = 299809225 \, m/s$$
$$v = 5.0124 \text{x} 10^{-16} \, m/s \tag{108}$$

Equation (108) indicates that the initial velocity is essentially 0 [m/s] relative to the speed of light to result in the value found in (107). This is something that has been confounding and requires further research.

15 Orbital Precession

A star system is a mechanical system. All mechanical systems adhere to energy conservation law. The total energy (H) for orbiting planets:

$$H = U + K$$

(109)

Where kinetic energy K is:

$$K = \frac{1}{2} m v^2$$

(110)

And potential energy U is:

$$U = \int_{\infty}^{x} F_g \, dx = \frac{-GMm}{x}$$

(111)

From centripetal force F_c we get velocity squared used in kinetic energy equation:

$$v^2 = \frac{F_c}{m} x$$

(112)

Combining (110) and (112):

$$K = \frac{1}{2} m \left(\frac{F_c}{m} x \right) = \frac{1}{2} F_c x$$

(113)

Where centripetal force is equal to the gravitational force:

$$F_c = F_g = \frac{GMm}{x^2}$$

(114)

Combining (113) and (114):

$$K = \frac{1}{2}\left(\frac{GMm}{x^2}\right)x = \frac{1}{2}\left(\frac{GMm}{x}\right)$$

(115)

Combining (109), (111) and (115):

$$H_m = \left(-\frac{GMm}{x}\right) + \left(\frac{1}{2}\frac{GMm}{x}\right)$$

(116)

Where the energy potential Λ of the planet mass m orbiting mass M at point x is:

$$V_m(x) = \left(\frac{1}{M}\right)H_m = \Lambda_m \quad [m^2/s^2]$$

$$\Lambda_m = -\frac{Gm}{x} + \frac{1}{2}\frac{Gm}{x}$$

(117)

$$\Lambda_m = -\frac{1}{2}\frac{Gm}{x}$$

Where scalar gravitational potential field is simply given by:

$$V(x) = \sum_{i=0}^{n} -\frac{Gm_i}{|x - x_i|}$$

(118)

The total potential sum of the system of two objects at their respective positions:

$$V_T = V_M(x_m) + V_m(x_M)$$

$$V_T = \left(-\frac{GM}{|x_m - x_M|}\right) + \left(-\frac{1}{2}\frac{Gm}{|x_M - x_m|}\right)$$

(119)

Where of course:

$$d = |x_m - x_M| = |x_M - x_m|$$

(120)

Therefore giving a final potential equation of:

$$V(d) = -\frac{GM}{d} - \frac{1}{2}\frac{Gm}{d} = -\frac{G\left(M + \frac{1}{2}m\right)}{d} \tag{121}$$

From the relative stationary frame of reference of object M, only the object m is moving, therefore applying the mass transform equation of (37), the following is derived:

$$V(d) = -\frac{G\left(M + \frac{1}{2}\left(\frac{m}{s(v)^2}\right)\right)}{d} \tag{122}$$

Where total gravitational acceleration is the sum of each individual gravitational acceleration at equal distance between mass objects, as in a two body system:

$$g = \frac{G\left(M + \frac{1}{2}\left(\frac{m}{s(v)^2}\right)\right)}{d^2} \tag{123}$$

Equation (123) gives the perihelion orbital precession result for Mercury of **43.6(+/- 0.4147) arc-secs per century** through its application in a computer simulation experiment, which is in agreement with observation and General Relativity.

Since the system of gravitational potential dynamic typically involves a much larger mass, the influence of the smaller mass is typically ignored, but as is seen here, neither potential energy nor kinetic energy must be ignored. This gives a different gravitational potential equation based on total mechanical energy H at point x in relation to small reference mass m_{ref}:

$$V(x) = \sum_{i=0}^{n} \Lambda_i(x) \tag{124}$$

Where Λ has a dimensional characteristic of v^2 which is interesting since potential is reduce to dimensions of length and time $L^2 T^2$, eluding to a more fundamental formulation for potential.

Figure 14: Plotting Mercury's first orbital track and 415 orbital tracks, 100 years later, using the framework of this book. As can be seen there is a slight orbital precession of 43.6(+/-0.4147)".

16 Gravitational Acceleration

The gravity field is an acceleration field mutually experienced between two or more objects. It can be described as electrodynamic pressure-differential between the quantum medium density between matter objects and the surrounding quantum medium density. The mechanics of this field are understood in General Relativity to behave as a curvature in space-time, where space-time is an abstraction of the more mechanistic dynamics of a quantum medium.

Classic gravitational acceleration is described as:

$$g = \frac{GM}{d^2}$$

(125)

As suggested by this research and the results obtained in regards to the perihelion precession of Mercury in the previous section, the following equation is considered more correct by accounting for kinetic energy and scaling:

$$g = \frac{G\left(M + \frac{1}{2}m\right)}{d^2}$$

(126)

The dimension to transform in this equation is the time component.

$$g = \frac{G\left(M + \frac{1}{2}m\right)}{d^2}\left(\tau(v)^2\right) \rightarrow a = \frac{k_e\left(Q + \frac{1}{2}q\right)}{d^2}$$

(127)

Equation (127) shows that the field of force (acceleration) increases in strength as the object's total kinetic velocity state increases, which is dependent on the natural terminal velocities (linear and rotational) induced by the surrounding quantum medium density.

17 Charge Type

Gravitational potential and kinetic energy are not the only dynamics at play. Thermodynamics play a significant role. Radiation pressure due to expelled energy quanta from a macroscopic object acts as a repulsive force upon colliding with another macroscopic object. Therefore a "cold" object that can absorb radiant energy more readily and experience less repulsion, compared to two "hot" objects equally expelling radiation of equal magnitude and frequency can (due to the same cause). It is reasoned that natural radiation, luminosity L, of an object is dictated by its type, size and mass-density which is subsequently determined by its velocity v:

$$L \propto \frac{1}{v} \tag{128}$$

This affects the overall gravitational potential equation. Defining this dynamic generally, equally massive objects repel each other, while objects of significant difference in size (mass-density) would attract. The following charge type coefficient equation between two objects is currently given as:

$$\delta = 2\sqrt{\left(\frac{(m_1 \rho_2 - m_2 \rho_1)}{(m_1 \rho_2 + m_2 \rho_1)}\right)^2 - 1} \tag{129}$$

And can be used as such:

$$g = \frac{\delta G\left(M + \frac{1}{2}\left(\frac{m}{s(v)^2}\right)\right)}{d^2} \tag{130}$$

There are some reservations about this formulation and subsequent use. A thermodynamic treatment with the application of cosmological luminosity effect on gravitational potential is preferred.

At the quantum scale, as at the macroscopic scale, charge type is considered a product, or classification, of radiation energy output ("heat") in relative reference to other objects. Some objects are "hot" and others are "cold", which can be classified as "positive" and "negative".

18 Photon Emission

Photon packets are absorbed by atoms increasing the orbital radii of the orbiting electrons, further increasing the size of the atom. This happens in a "hot" environment containing many transient photons colliding with atoms. When the environment "cools", the absorbed photons are emitted back out of the atom, which reduces the orbital radii of the electrons decreasing the overall size of the atom.

At the macroscopic scale, a change in the orbital radii of a gas-giant planet would be the same as an electron changing orbital energy levels in an atom.

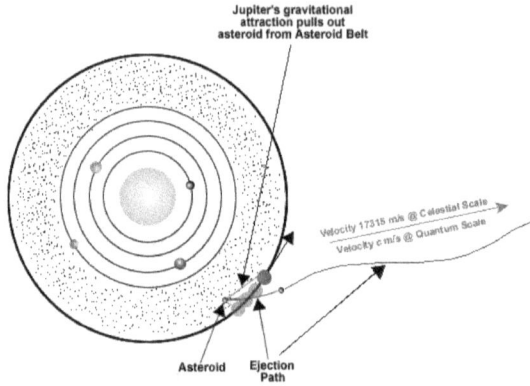

Figure 15: Photon ejection was electron/gas-giant changes orbit

In the Bohr model using hydrogen, this is defined by this equation:

$$E_{photon} = hf = E_i - E_f = \frac{k_e e^2}{2}\left(\frac{1}{r_f} - \frac{1}{r_i}\right)$$

(131)

This can be changed to an equivalent macroscopic version using the understanding orbital planet masses, at least in our Solar System, are not the same:

$$E_{macro\,photon} = E_i - E_f = \frac{G\,Mm}{2}\left(\frac{1}{r_f} - \frac{1}{r_i}\right)$$

(132)

This macroscopic photon is reasoned to change from a cloud of space dust to the group of asteroids.

The two primary orbital regions (bands) of the gas-giant in the Solar System are Jupiter and Saturn in the closer region and Uranus and Neptune in the further region. This corresponds to the Beryllium's shell configuration:

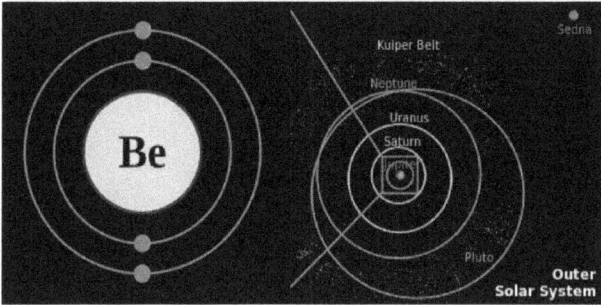

Figure 16: Beryllium electron shell configuration. Beryllium image is original work by Greg Robson (created by Pumbaa on Wikipedia).

The Asteroid Belt and Jupiter's Lagrange point regions in space have a buildup of asteroid and space dust. This is reasoned to be a repository of macroscopic photonic material, and representative of the macroscopic energy of the star system, akin to the energy level of the atomic system.

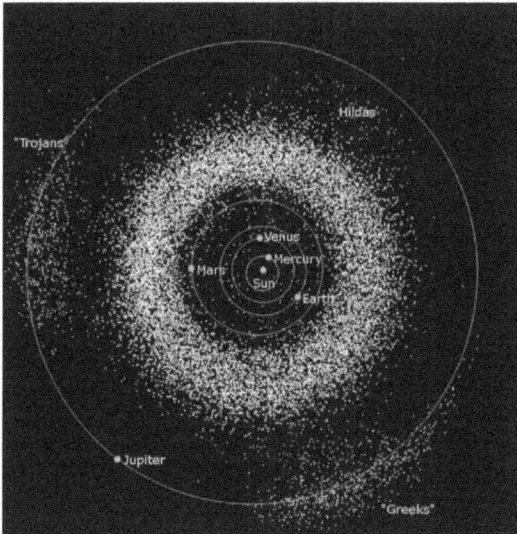

Figure 17: Asteroids in star system are repository of photons and representative of system's macroscopic energy level

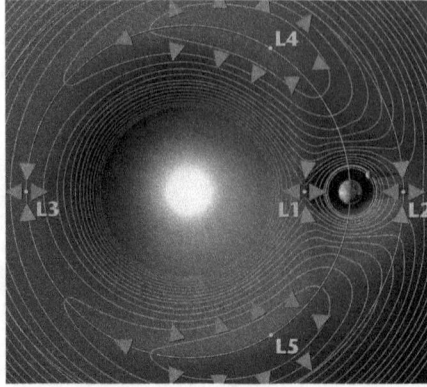

Figure 18: Jupiter Lagrange points regions of collected asteroids and space dust

Using this understanding, we test equation (132) by having Uranus fall into Saturn's orbit:

$$E_{macro\ ph} = \frac{G\,M_{Sat}\,m_U}{2}\left(\frac{1}{r_U} - \frac{1}{r_{Sat}}\right)$$

$$E_{macro\ ph} = 5.7933\text{x}10^{29}[J]$$

(133)

Now a quantum particle energy can be estimated by de Broglie's matter wave relations:

$$E_{macro\ ph} = h_s\,f = m\,c_s^2$$

(134)

Where h_s and and c_s are the macroscopic versions of Planck constant and speed of light. The macroscopic version of Planck constant is yet unknown conclusively. Macroscopic version of the speed of light is strongly reasoned to be 17314.56 [m/s]. This gives the photon a macroscopic scaled mass and quantum charge respectively:

$$m_{m.ph} = \frac{E_{m.ph}}{c_s^2} = 6.44\text{x}10^{12}[kg = C]$$

$$q_{ph} = \frac{m_{m.ph}}{S^2} = 5.3\text{x}10^{-34}[C = kg]$$

(135)

This macroscopic mass of an macroscopic photon, or asteroid, is close in value to the mass of the Comet Showmaker-Levy 9 that struck Jupiter in 1994 at an estimated mass of $7.136\text{x}10^{11}$ [kg][7].

When using Neptune's orbit, the values are:

$$m_{m.ph}=\frac{E_{m.ph}}{c_s^2}=1.033\times10^{13}[kg=C]$$

$$q_{ph}=\frac{m_{m.ph}}{S^2}=8.4956\times10^{-34}[C=kg]$$

(136)

Numerically these charge values are remarkably close to the numerical value of Planck's constant. Given this framework, it's something not to dismiss so easily. *Their average value is 6.8824x10^{-34} which is extremely close to Planck's constant.* This may mean that Planck's constant is nothing more than the charge value (quantum mass in relation to sub-quantum medium) of a quantum photon, and frequency of a photon is its kinetic velocity squared (rotational and linear). It's definitely something to explore.

Using Jupiter, the following values for Uranus are derived:

$$m_{m.ph}=5.7129\times10^{13}[kg=C]$$

$$q_{ph}=\frac{m_{m.ph}}{S^2}=4.7\times10^{-33}[C=kg]$$

(137)

And the following values for Neptune are derived:

$$m_{m.ph}=7.6462\times10^{13}[kg=C]$$

$$q_{ph}=\frac{m_{m.ph}}{S^2}=6.2905\times10^{-33}[C=kg]$$

(138)

Where the ratio between the averages of (137) and (138) and 6.8824x10^{-34} is 7.9824, almost the whole number 8 which is interesting because there are 8 known orbits in our Solar System, 4 inner and 4 outer, though this maybe mere coincidence. Using the derived energy-mass difference between Jupiter and Saturn, two objects in the same "shell", the ratio in charge value using the lowest average value of 6.8824x10^{-34} is 27.8899. Again this is almost another whole number.

Considering the increased probability that Planck's constant is possibly the charge (quantum mass) of a photon ([kg=C]), then frequency, through dimensional analysis, is related to the combination of linear and rotational velocity, but where linear velocity is restricted to the speed of light c leaving only rotational velocity variable.

$$h \rightarrow [kg.m/s].[m]$$
$$f \rightarrow [1/s]$$
$$m \rightarrow [kg] \qquad (139)$$
$$v^2 \rightarrow [m/s][m/s]$$

Notice that h has a velocity component, and given its constant nature, this velocity component must be constants also. This velocity is reasoned to be c. Frequency f has a single time component which is variable. This time component in combination with h's remaining length component gives an additional variable velocity. Two velocities, but where only one is variable.

This analysis forms relation:

$$E_{photon} = hf$$
$$h \rightarrow q_{photon} \quad [constant]$$
$$f \rightarrow \sqrt{v_{linear}^2 + v_{rotation}^2} = \sqrt{c^2 + v_{rotation}^2} \qquad (140)$$
$$E_{photon} \rightarrow (q_{photon})(c^2 + v_{rotation}^2)$$

If relation (140) is true, then high frequency quantum particles definitely have higher *total* kinetic energy in reference to the sub-quantum medium.

19 Particle Wave Duality

The classical wave equation is:

$$\frac{\partial E_\psi}{\partial x^2} = \frac{1}{c^2} \frac{\partial E_\psi}{\partial t^2}$$

(141)

Where wave energy is:

$$E_\psi = \iint \left(\frac{1}{c^2} \frac{\partial E_\psi}{\partial t^2} \right) \partial x \, \partial x$$

(142)

From which E_ψ is defined in relation to kinetic energy:

$$E_\psi = \frac{1}{2} m \left(\frac{x}{t} \right)^2$$

(143)

Deriving the second order time derivative:

$$\left(\frac{1}{c^2} \right) \frac{\partial^2}{\partial t^2} \left(\frac{mx^2}{2} \right) t^{-2} = \frac{3 \, mx^2}{c^2 t^4}$$

(144)

Finally concluding in:

$$E_\psi = \frac{3 \, mx^2}{c^2 t^4} \iint \left(x^2 \right) \partial x \, \partial x = \frac{1}{4} \left(\frac{m}{c^2} \right) \left(\frac{x^4}{t^4} \right) = \frac{1}{4} m \left(\frac{v^4}{c^2} \right)$$

(145)

If $v = c$ then equation (145) becomes:

$$E_\psi = \frac{1}{4} m c^2$$

(146)

Since de Broglie related the following for a photon:

$$E_\lambda = hf = mc^2$$

(147)

Then

$$E_{net} = E_\lambda - E_\psi = \left(\frac{4}{4} - \frac{1}{4} \right) mc^2 = \frac{3}{4} m c^2 \tag{148}$$

Where mass for this net energy equates to:

$$m = \frac{4}{3} \frac{E}{c^2} \tag{149}$$

Which is electromagnetic mass defined measured in classical quantum physics. This means equation (146) is the energy of Henri *Poincaré's stresses* (from quantum medium) of non-electromagnetic energy E_p of a wave [4][5][6].

The significance of this derivation is that it shows that a quantum particle consists of an electromagnetic mass under non-electromagnetic stresses of a wave through the quantum medium. Both energy components together form the particle's total energy. This describes wave-particle duality, with the help of the following diagram:

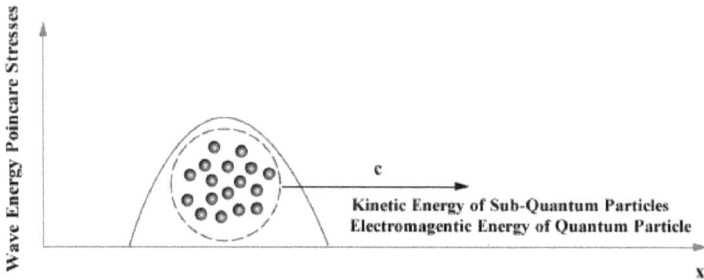

Figure 19: Poincare stresses quantum wave energy

Quantum particles arise when wave energy in the quantum medium is sufficient enough to constitute a low charge-density boson from condensed sub-quantum particles in the wave crest, or high enough to constitute high charge-density fermion particle.

The immense increase in the passage of time, factor *T*, also contributes to wave-particle duality as the physical state of the quantum particle will be in constant physical flux from the macroscopic time-frame of reference.

20 Frequency Shift Refraction

It is also known that the velocity of light is defined by artifacts of the medium it propagates through as described by this equation:

$$v = \sqrt{\frac{1}{\mu_f \varepsilon_f}} = c = \sqrt{\frac{P}{\rho_f}} \tag{150}$$

Where the following is derived from drag force counter-balanced by the electric force:

$$\mu_f \varepsilon_f = \rho_f \left(\frac{C_d A}{2qE} \right) \tag{151}$$

Variable μ is vacuum permeability, ε is vacuum permittivity, P is bulk elasticity modulus and ρ_f is the charge-density of the quantum medium (quantum mass-density). Vacuum permeability and permittivity values are assumed constants in regards to the vacuum of space.

If the quantum medium density can vary as ISM density does.

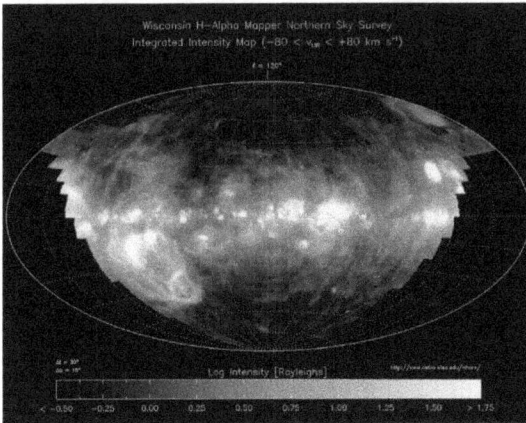

Figure 20: ISM showing density distribution directly sugges-tion that the relatively scaled quantum medium also varies in density from region to region. (The Wisconsin H-Alpha Mapper is funded by the (US) National Science Foundation)

As photon's travel through varying quantum medium densities, they experience a form of incident refraction which changes their direction of motion, gradually if the quantum medium density change is gradual, and their initial relative quantum mass (charge) directly affecting their energy and frequency.

Refraction index is inversely proportional to velocity:

$$\frac{v_1}{v_2} = \frac{n_2}{n_1} \tag{152}$$

The speed of light is represented by (150), and can be modified into the following representation:

$$\frac{c_1}{c_2} = \frac{\sqrt{\dfrac{1}{\mu_1 \varepsilon_1}}}{\sqrt{\dfrac{1}{\mu_2 \varepsilon_2}}} = \frac{\sqrt{\mu_2 \varepsilon_2}}{\sqrt{\mu_1 \varepsilon_1}} = \frac{n_2}{n_1} \tag{153}$$

This indicates that the refraction index is proportional to the permittivity and permeability of the respective space region (note by numeric subscript). In equation (151), permittivity and permeability are directly proportional to the density of the quantum medium ρ_f which means that if all other parameters are constant on the left side of equation (154) except for ρ_f, than permittivity and permeability constants (variables) will change. Based on the scale relation between ISM and the quantum medium, it is reasoned that as the quantum medium varies in density from region to region, so too do permittivity and permeability constants change for vacuum space from region to region. This has direct consequences on the velocity of light (or terminal velocity) in those regions and when they move from one density of the quantum medium to another.

$$\mu_f \varepsilon_f = \rho_f \left(\frac{C_d A}{2qE} \right) \tag{154}$$

The energy of the photon is as so:

$$E = hf = \frac{1}{2} m_q c^2$$
$$f = \frac{m_q c^2}{2h} \tag{155}$$

Where m_q is its inertial mass, which is virtually non-existent due to its very small size, and the way it propagates through the quantum medium via a temporal dependent, physical compression wave. It is important to note that the photon charge is reasoned to exist, according to the scaling framework of this document, but is significantly small enough to be virtually undetectable.

Substituting (154) into (155) we get:

$$f = \frac{m_q \left(\sqrt{\frac{1}{\mu \varepsilon}} \right)^2}{2h} = \frac{m_q}{2h \mu \varepsilon} \tag{156}$$

The following image presents a photon traversing between two different densities of quantum medium (QM).

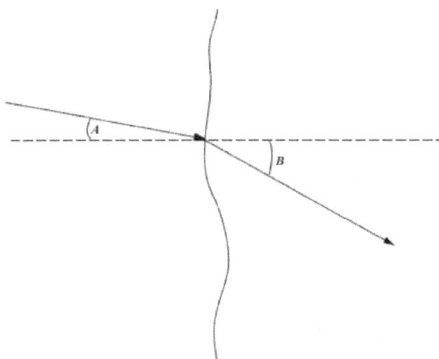

Figure 21: Refraction of light indicates a frequency change
also. Refraction of light in the vacuum of space can be understood as curvature of light due to gravity which is directly
related to the gradient density variation in the density of the
quantum medium around macroscopic matter objects

The frequency shift can be calculated as such:

$$\Delta f = f'' - f'$$
$$\Delta f = \frac{m_q}{2h \mu_2 \varepsilon_2} - \frac{m_q}{2h \mu_1 \varepsilon_1} \tag{157}$$
$$\Delta f = \frac{m_q}{2h} \left(c_2^2 - c_1^2 \right)$$

Where is can be further elaborated as:

$$\Delta f = \frac{m_q}{2h\left(\dfrac{\rho_2 C_{d2} A}{2qE}\right)} - \frac{m_q}{2h\left(\dfrac{\rho_1 C_{d1} A}{2qE}\right)}$$

$$\Delta f = \frac{m_q q E}{h\left(\rho_2 C_{d2} A\right)} - \frac{m_q q E}{h\left(\rho_1 C_{d1} A\right)} \tag{158}$$

$$\Delta f = \frac{m_q q E}{hA}\left(\frac{1}{\rho_2 C_{d2}} - \frac{1}{\rho_1 C_{d1}}\right)$$

From this equation, based on the frequency shift, the density and the coefficient value of the quantum medium can be partially extrapolated, or a relative proportional value can be extrapolated, using the length and mass transformation equations of this framework.

20.1 Red and Blue Shift

What equations (152) to (158) offer is an alternative to the Doppler frequency shift explanation for Red Shift and Blue Shift in the frequency of light. This alternative does not require star systems to be moving away, or towards, our Solar System to explain these shifts. In turn, neither does this alternative explicitly mean that the Doppler frequency shift of light, due to the movement of stars, does not also exist. The most prudent reasoning would be to state that both effects in congruency are a possibility given other observational data.

This has grave implications on the Big Bang theory, though it does permit the existence of a type of Big Bang event. The framework of this document indicates that if the quantum medium were reduced considerably in density, or a matter object were reduced in velocity to 0 m/s relative to a Universal static frame of reference (like the SQM), that the matter object would expand to fill that region absorbing material from the collective sub-quantum mediums. The collective sub-quantum medium would be less affected by such an explosion. An extremely large catastrophic explosion could in effect scatter the density of the quantum medium so drastically resulting in this particular type of Big Bang event.

21 Future Research

This research is only at the threshold of potentially more revealing physical relations and perplexing possibilities.

21.1 The Big Bang

The framework can be expanded more speculatively to encompass the Big Bang theory through the theory of Universal inflation theory. Inflation theory is the asymmetric expansion of Universe. This indicates that the Universe may have started as a matter particle, possibly a quantum particle, which slowed down somehow kinetically and began to grow by absorbing surrounding and per-existing, or pre-"Big Bang", quantum medium. It is possible that Universal expansion is truly due to an explosion, one which we are still in its initial phase. Coupling the birth of the Universe with this framework may reconcile many disparaging details of the Big Bang and form a larger picture of the exo-Universe.

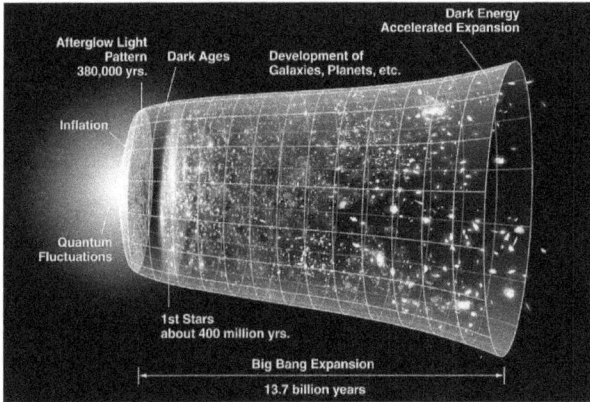

Figure 22: Big Bang explained as aspect of fractal scaling theory

The radius of the observable Universe is 4.3235×10^{26} [m]. Applying length transform using discrete scale factor S is:

$$r_o = \frac{r_U}{S} = 3921.54 \; [m] \tag{159}$$

This value is close to the radius of Mars, a small rock planet. Using this framework, the observable Universe could merely be a super-cosmological planet; one of an unknown many.

21.2 Topology

Within the last year (2012), an emerging pattern has been arising in the research to reconcile quantum physics with astrophysics. The pattern is topological and is described in more detail in the book GPRA:REPMES[1]. Currently this topology is bsaed on potentials.

The rest gravitational potential energy formulation containing a wave formation which closely matches classical Newtonian gravitational potential energy and derivative force slope is the following equation. This equation (160) adheres to formula $k = 2/radius$. This formula is most certainly incomplete, but for the sake of presentation and argument, it is included.

$$U(x)=-G\,m_{ref}\left(\frac{M}{x}\right)\sin\left(G\,m_{ref}\left(\frac{M}{\left(\frac{1}{2}r_M\right)x}\right)\left(\frac{1}{F_{ref}()}\right)\right)$$

$$U(x)=-\left(\frac{G(1\,kg)(5.98\times10^{24}\,kg)}{x}\right)$$

$$\sin\left(\frac{G(1\,kg)(5.98\times10^{24}\,kg)}{\frac{1}{2}(6.378\times10^6\,m)x}\left(\frac{1}{F_{ref}()}\right)\right)$$

(160)

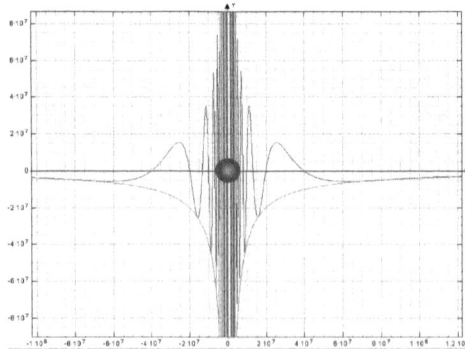

Figure 23: 2D Potential Energy Wave Density of Earth

Where $F_{ref}()$ is a reference force variable for dimensional correction and has the value of 1 [kg.m.s^{-2}] leaving dimensional distance, or meters (*m*), in the sin-wave function representing rads which is then nullified by the wavelength or presented by the radius of the planet (Sun). Again, this high level fractal analysis, defining a fractal topological definition of gravity, is still a work in progress but is interesting.

Now it must be mentioned that at periodic intervals in this model, there would have to exist concentric spherical intervals of very low density even within the planet's body. Low density of material existence at the crust of the Earth is of heavier density than its immediate atmosphere. In the separation space from very dense rock on the surface to that of less dense air exists a virtual membrane of electromagnetic shielding where few atoms exist. **This also means that when the crest and dips of the topological gravitational potential wave are tight together, they essentially present the formation of macroscopic matter as macroscopic matter contains more condensed energy than the low density quantum medium.** *Essentially concentrated quantum medium forms quantum particles and macroscopic matter.*

As described, the dips and rises in the potential energy wave pattern could represent the theorized concentric rings of Earth's internal structure, along with its levels of atmosphere. This would also indicate that at certain distances internally in planets, there are "cooler" concentric rings and very hot" concentric rings.

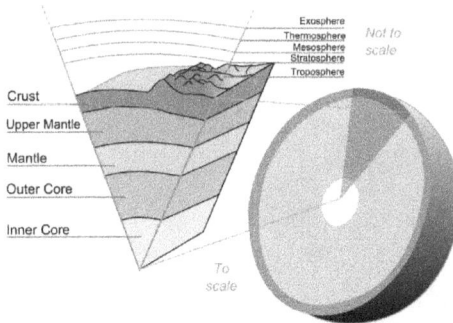

Figure 24: Concentric Interior of Rock Planets. Based on the public domain image Earth-crust-cutaway-english.png by Jeremy Kemp (created by Surachit on Wikipedia).

Of course this topological mathematical model needs more work, but consider the following equation (161), a variation of equation (160), and resulting Figure (25) of the this topological wave theory of system potentials, due to density variations in the quantum medium (QM), of the Sun.

$$\Psi(x) = A \sin(kA)$$

$$k = \frac{2\pi}{\lambda} = \frac{2\pi}{r_{radius}}$$

$$U(x) = -\left| \left(G\, m_{ref} \frac{M_{sun}}{x} \right) \sin\left(\left(\frac{2\pi}{r_{sun}} \right) \left(G\, m_{ref} \frac{M_{sun}}{x} \right) \left(\frac{1}{F_{ref}()} \right) \right) \right| \tag{161}$$

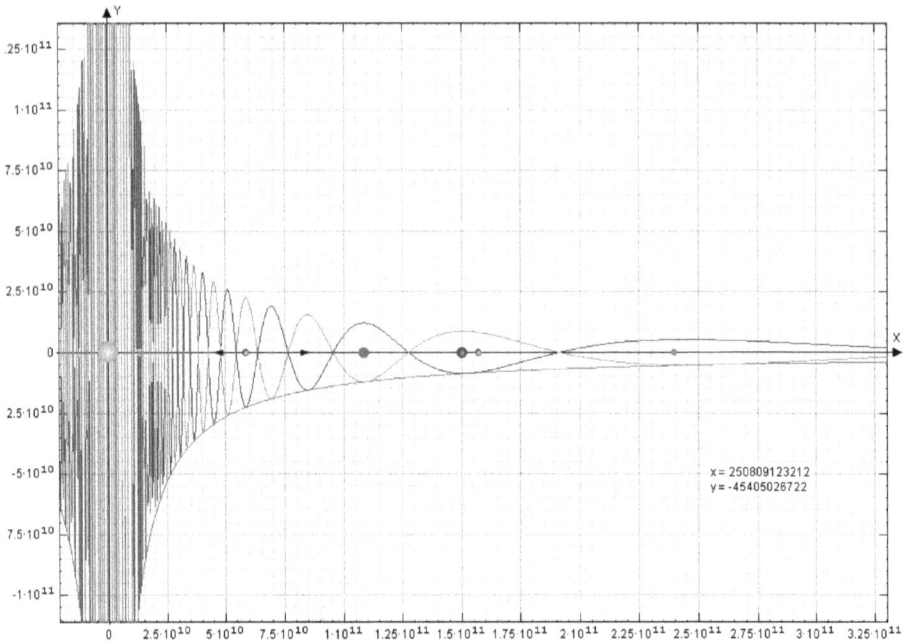

x= 250809123212
y= -45405026722

Figure 25: Sun's Topological Potential Energy Wave Representation with Inner Planets

Figure (25) wave amplitude is multiplied by 10 for illustrative visibility, **but the distance is correctly scaled and shows the inner planets at their exact orbital distances. This gives some validity to equation (161).** This is definitely an equation that needs further exploration. Planet orbital radii are available in the Appendix (22.2).

The pockets in the wave pattern can be directly considered pockets of varying quantum medium (QM) density, or pockets of less quantum medium density, in relation to the planet's themselves and the surrounding space. The closer the dips and rises, and the higher they are, the more compact and energetic the quantum medium is, until it actually forms the material Sun itself. The same formulation can be applied to the planets. The further apart the top wave crest is from its respective bottom mirrored wave crest, the blue and green waves in Figure (25), the less dense the quantum medium is (*hotter*). The flatter the distance between the two, the more dense the quantum medium is (*colder*).

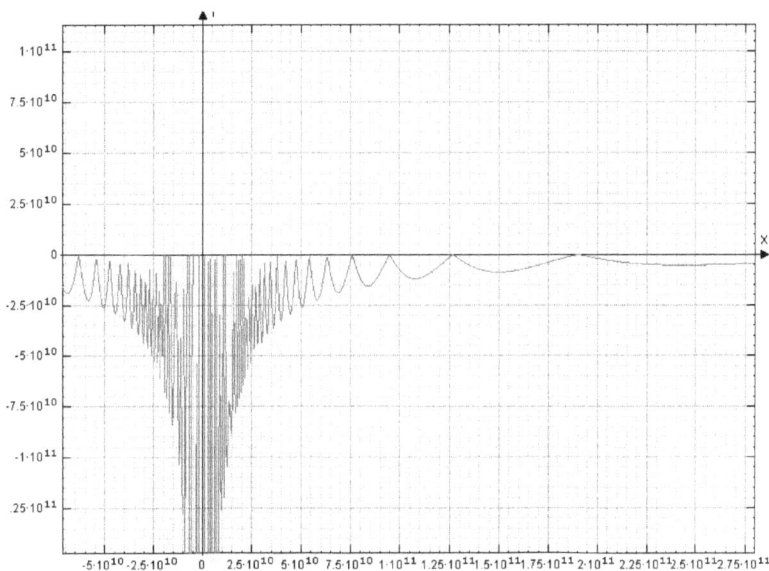

Figure 26: Sun's Topological Potential Energy Wave Representation with Inner Planets (x10 for visibility)

In this model, planets take the path of least resistance, regions of less dense quantum medium, and form stable orbits in their respective pockets; with the exclusion of Mercury. The density can be measured by the distance between the two waves on the y-axis. The further the distance, the stronger the quantum medium distortion and the less quantum medium density (*hotter*) in that region. The closer the two waves are in the y-axis, the denser the quantum medium (*colder*) and the more inertia a matter object would have (the more mass it would have). This falls inline with the scaling framework of this document. Mass increase and decrease is related to velocity and the respective effects of the quantum medium on fast and slow objects. These regions, or pockets, of less dense quantum medium allow objects within them to travel at faster velocities than the denser quantum medium in deep space permits, further away from the Sun. It is clear that the density energy, or potential energy density, of the quantum medium (QM) is essentially infinitesimal compared to energy density of a matter object and that of the Sun as presented in the prior figure, which the gravitational G constant also indicates. The quantum medium (QM) plays the "yin" to matter's "yang". The pockets in which these planets reside in are not absolute boundaries and can definitely be penetrated (quite easily) as Mercury's orbit is very elliptical and passes through boundaries beyond its median orbit. They are merely regions of space containing less quantum medium (*QM*) due to the causes of the Sun's own expelled energy quanta, which travel and superposition in a wave pattern. This wave pattern is further due to the expelled energy quantum of the planets themselves and the orbital motion of the planets. As a planet moves near these wave boundaries, the quantum medium density increases and opposes the motion of the planets by increased resistance, thus increased inertial mass of the planet, yet if the object's velocity is increased, the planet's distortion of the quantum medium (ex. magnetic field) will act as a shield and make the planet essentially more "aerodynamic" in this medium as the bulk of the medium flows with the displaced medium along the distortion boundary (magnetic field).

Here is a 3D plot of two objects using a scaled version of equation (161). Here we're using the formulation for density $\rho = f(x, y, z)$. to change the visual opacity in that region.

Figure 27: 3D Quantum Medium Density Wave Formation
Representation.

Figure (27) is interesting for the fact that two objects rendered are *solely* a result of a 3D plot of wave pattern energy potentials in that space only. It clearly aids the argument that macroscopic matter objects, and quantum particles, are a topological wave artifacts of the quantum medium.

21.3 Parallel Worlds

This framework predicts the Universe is infinitely sized and fractal (through scaling). This would greatly increase the probability of parallel worlds existing, separated by distance, variable time frames of reference and scale. This would also increase the probability of *relative* time travel between them, between identical worlds at different stages of development. This would also bring into question the definition of unique self in a super-fractal-Universe context with the existence of parallel worlds. If an unknown number of parallel copies of us exist in a super-fractal-Universe, than do our parallel selves collectively constitute our existence? Are our actions predetermined or our we unconsciously clairvoyantly entangled to the thoughts and actions of our parallel selves who have already experienced our *relative* future? If so, then we will never cease to exist somewhere in the Universe, separated by distance, passage of time and scale. The possibilities are perplexing.

This theory would also increase the probability that all life in the Universe is essentially one entity through a cyclic treatment of variation between almost identical worlds. An example would be that you and your neighbour are the same person through a twist on dimensional relativity. The following diagram depicts this almost parallel world cyclic theory, as our parallel selves on each almost identical parallel world varies in difference by only an arbitrary 5%. If the connection of almost parallel worlds cycle back to the originating world through 20 parallel worlds, this would result in 100% change from the originating person back on the originating world, and living concurrently with the originating person. It's wonderfully perplexing.

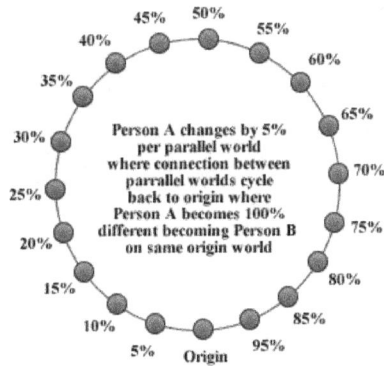

28. Figure: Almost identical parallel world cycle back theory, where you and your neighbour are inter-dimensionally the same person.

Science has merely scratched the surface of what we don't know, and we don't know a lot. We must cast aside ignorant, dogmatic, and prideful thoughts of contemporary knowledge and focus on testing every possibility.

22 Appendix

22.1 Constants

Constant	Symbol	Value
acceleration due to gravity	g	9.8 m s-2
atomic mass unit	amu, mu or u	1.66E-27 kg
Avogadro's Number	N	6.022E23 mol-1
Bohr radius	a0	0.529E-10 m
Boltzmann constant	k	1.38E-23 J K-1
electron charge to mass ratio	-e/me	-1.7588E11 C kg-1
electron classical radius	re	2.818E-15 m
electron mass energy (J)	mec2	8.187E-14 J
electron mass energy (MeV)	mec2	0.511 MeV
electron rest mass	me	9.109E-31 kg
Faraday constant	F	9.649E4 C mol-1
fine-structure constant	α	7.2970E-03
gas constant	R	8.314 J mol-1 K-1
gravitational constant	G	6.67E-11 Nm2kg-2
neutron mass energy (J)	mnc2	1.505E-10 J
neutron mass energy (MeV)	mnc2	939.565 MeV
neutron rest mass	mn	1.675E-27 kg
neutron-electron mass ratio	mn/me	1838.68
neutron-proton mass ratio	mn/mp	1.0014
permeability of a vacuum	μ0	4πE-7 N A-2
permittivity of a vacuum	ε0	8.854E-12 F m-1
Planck constant	h	6.626E-34 J s
proton mass energy (J)	mpc2	1.503E-10 J
proton mass energy (MeV)	mpc2	938.272 MeV
proton rest mass	mp	1.6726E-27 kg
proton-electron mass ratio	mp/me	1836.15
Rydberg constant	r∞	1.0974E7 m-1
speed of light in vacuum	C	2.9979E8 m/s

22.2 Planet Properties

object	code	type	equatorial radius m	equatorial radius Earth=1	mass kg	mass Earth=1	average density Kg/m^3	surface gravity Earth=1	escape velocity km/s	rotation velocity m/s	inclination of equator to orbit	semimajor axis AU=149597871 km AU	semimajor axis AU=149597871 km m	Aphelion AU	Perihelion AU	orbital period yr	orbital period days	average orbital velocity m/s	orbital eccentricit y	orbital inclination to eliptic
Sun	Su	star	6.9634E+08	109.177	1.9891E+30	333059.84	1406.37	27.94	617.7	1996.94								220000		
Mercury	Me	rock	2.4428E+06	0.383	3.3026E+23	0.0553	5408.79	0.378	4.4	3.026	0°05'	0.3871	5.7909E+10	0.466 697	0.307 499	0.24084	87.968	47870	0.2056	7°00'18'
Venus	V	rock	6.0528E+06	0.949	4.8673E+24	0.815	5239.98	0.907	10.4	1.81	177°18'	0.7233	1.0820E+11	0.728 213	0.718 440	0.61519	224.702	35020	0.0068	3°23'40'
Earth	E	rock	6.3781E+06	1	5.9722E+24	1	5495.04	1	11.2	465.1	23°26'	1	1.4960E+11	1.01671388	1.00000261	1	365.256	29780	0.0167	0°00'00'
Moon	Em	rock	1.7380E+06	0.2725	7.3477E+22	0.0123032	3341.09		2.38	4.627	18.29-28.58°	0.00257	3.8447E+08	0.0027	0.0024	0.0748536	27.321582	1022	0.0549	5.145°
Mars	Mr	rock	3.3931E+06	0.532	6.4141E+23	0.1074	3919.59	0.377	5	241.17	25°11'	1.5237	2.2794E+11	1.665861	1.381497	1.88081	686.978	24120	0.0934	1°50'59'
Jupiter	J	gas	7.1396E+07	11.194	1.8981E+27	317.8284	1245.11	2.364	59.5	12600	3°07'	5.2028	7.7833E+11	5.458104	4.950429	11.8616	4332.53	13070	0.0485	1°18'12'
Saturn	Sa	gas	6.0330E+07	9.459	5.6832E+26	95.1609	617.87	1.064	35.5	9870	26°44'	9.5428	1.4276E+12	10.11596	9.048076	29.4609	10760.78	9650	0.0555	2°29'20'
Uranus	U	gas	2.6201E+07	4.108	8.6810E+25	14.5357	1152.17	0.889	21.3	2590	97°52'	19.1921	2.8711E+12	20.083305	18.375518	84.0241	30690.34	6800	0.0463	0°46'24'
Neptune	N	gas	2.5225E+07	3.955	1.0241E+26	17.1478	1523.14	1.125	23.5	2680	29°34'	30.0669	4.4982E+12	30.44125	29.76607	164.7734	60184.5	5440	0.009	1°46'12'
Pluto	P	ice	1.1672E+06	0.183	1.2542E+22	0.0021	1882.94	0.067	1.3	13.1055	122°	39.4817	5.9064E+12	48.871	29.657	247.92	90554	4740	0.2488	17°08'30'

23 References

[1] DeMelo, R. L. (2012). *"Relativity and the Electrodynamics on the Phenomena of Matter and Energy through Scale (GPRA:REPMES)"*. Toronto, www.gpofr.com, Library and Archives Canada (LAC), ISBN 978-0-9810242-8-8. *(contains 160 references)*

[2] Turok, N. (2012). *"The Universe Within : From Quantum to Cosmos"*. Toronto, House of Anansi Press Inc., Library and Archives Canada (LAC), ISBN 978-1-77089-015-2.

[3] Palmer, T. N. (2009). *"The Invariant Set Postulate: a new geometric framework for the foundations of quantum theory and the role played by gravity"*. Proceedings of the Royal Society a Mathematical Physical and Engineering Sciences 465 (2110): 3165. arXiv:0812.1148. doi:10.1098/rspa.2009.0080

[4] Poincaré, Henri (1906), *"The End of Matter"*, Athenæum

[5] Poincaré, Henri (1905), *"On the Dynamics of the Electron"*, Comptes Rendus 140: 1504–1508.

[6] Poincaré, Henri (1906), *"On the Dynamics of the Electron"*, Rendiconti del Circolo matematico di Palermo 21: 129–176, doi:10.1007/BF03013466

[7] Crawford, D. A.,*"Comet Shoemaker-Levy 9 Fragment Size and Mass Estimates from Light Flux Observations"*, Sandia National Laboratories, http://www.lpi.usra.edu/meetings/lpsc97/pdf/1351.PDF

[8] French,A.P., Taylor,E.F. (1978). *"An Introduction to Quantum Physics (The MIT Introductory Physics Series)"*, MIT, W.W.Norton & Comany Inc. ISBN: 0-393-09106-0.

[9] Lu, D (2003). *"University Physics (Second Edition)"*. Beijing, Higher Education Press

[10] Wegner, J.L., Haddow, J.B. (2009). *"Elements of Continuum Mechanics and Thermodynamics"*. Cambridge University Press. ISBN: 978-0-521-86632-3.

[11] Steven, H.S. (1994). *"Nonlinear Dynamics and Chaos: With Applications to Physics, Biology, Chemistry and Engineering"*. Perseus Books Publishing. ISBN: 978-0-7382-0453-6.

[12] O. Heaviside (1893). *"A gravitational and electromagnetic analogy"*. The Electrician 31: 81–82.

www.ingramcontent.com/pod-product-compliance
Lightning Source LLC
Chambersburg PA
CBHW031952190326
41519CB00007B/775